WORKBOOK to Accompany

Residential Construction Academy

Plumbing

SECOND EDITION

KEVIN STANDIFORD

Australia • Brazil • Japan • Korea • Mexico • Singapore • Spain • United Kingdom • United States

**Workbook to Accompany
Residential Construction Academy:
Plumbing, Second Edition**
Kevin Standiford

Vice President, Career and Professional Editorial: Dave Garza

Director of Learning Solutions: Sandy Clark

Senior Acquisitions Editor: Jimes DeVoe

Managing Editor: Larry Main

Product Manager: Brooke Wilson, Ohlinger Publishing Services

Editorial Assistant: Cris Savino

Vice President, Career and Professional Marketing: Jennifer Baker

Marketing Director: Deborah Yarnell

Marketing Manager: Katie Hall

Production Director: Wendy Troeger

Production Manager: Mark Bernard

Content Project Manager: David E. Plagenza

Art Director: Casey Kirchmayer

Technology Project Manager: Joe Pliss

© 2012 Delmar, Cengage Learning

ALL RIGHTS RESERVED. No part of this work covered by the copyright herein may be reproduced, transmitted, stored or used in any form or by any means graphic, electronic, or mechanical, including but not limited to photocopying, recording, scanning, digitizing, taping, Web distribution, information networks, or information storage and retrieval systems, except as permitted under Section 107 or 108 of the 1976 United States Copyright Act, without the prior written permission of the publisher.

> For product information and technology assistance, contact us at
> **Cengage Learning Customer & Sales Support, 1-800-354-9706**
> For permission to use material from this text or product,
> submit all requests online at **www.cengage.com/permissions**
> Further permissions questions can be e-mailed to
> **permissionrequest@cengage.com**

Library of Congress Control Number: 2010936152

ISBN-13: 978-1-111-30780-6

ISBN-10: 1-111-30780-6

Delmar
Executive Woods
5 Maxwell Drive
Clifton Park, NY 12065
USA

Cengage Learning is a leading provider of customized learning solutions with office locations around the globe, including Singapore, the United Kingdom, Australia, Mexico, Brazil, and Japan. Locate your local office at **www.cengage.com/global**

To learn more about Delmar, visit **www.cengage.com/delmar**

Purchase any of our products at your local bookstore or at our preferred online store **www.cengagebrain.com**

Notice to the Reader
Publisher does not warrant or guarantee any of the products described herein or perform any independent analysis in connection with any of the product information contained herein. Publisher does not assume, and expressly disclaims, any obligation to obtain and include information other than that provided to it by the manufacturer. The reader is expressly warned to consider and adopt all safety precautions that might be indicated by the activities described herein and to avoid all potential hazards. By following the instructions contained herein, the reader willingly assumes all risks in connection with such instructions. The publisher makes no representations or warranties of any kind, including but not limited to, the warranties of fitness for particular purpose or merchantability, nor are any such representations implied with respect to the material set forth herein, and the publisher takes no responsibility with respect to such material. The publisher shall not be liable for any special, consequential, or exemplary damages resulting, in whole or part, from the readers' use of, or reliance upon, this material.

Printed at CLDPC, USA, 03-19

Table of Contents

Preface xiv

CHAPTER 1
Plumber's Toolbox 1

   Objectives 1
      Knowledge-Based 1
      Skill-Based 1
   Keywords 1
   Introduction 1
      Hand Tools 1
      Levels 1
      Measuring 2
      Tape Measures 2
      Squares 2
      Screwdrivers 2
      Pliers 2
      Adjustable Wrenches 3
   Chapter Review Questions and Exercises 3
   Job Sheet 1: The Plumber's Toolbox 5
      Instructor's Response 6
   Job Sheet 2: The Plumber's Toolbox 7
      Instructor's Response 8
   Job Sheet 3: Hand Tools 9
      Instructor's Response 10
   Job Sheet 4: Using a Level to Set the Slope of a Pipe 11
      Instructor's Response 12

Job Sheet 5: First Aid Kits 13
    Instructor's Response 14

CHAPTER 2
Power Tools 15

Objectives 15
    Knowledge-Based 15
    Skill-Based 15
Keywords 15
Introduction 15
    Electrical Safety 15
    Ladder Safety 16
    Fall Protection 16
    Ear Protection 16
    Drills 16
    Right angle Drill 16
    Pistol Drill 16
    Hole Hawg 17
Chapter Review Questions and Exercises 17
Job Sheet 1: Power Tool Checklist 19
    Instructor's Response 20
Job Sheet 2: Right Angle and Hole Hawg Safety 21
    Instructor's Response 22

CHAPTER 3
Types of Pipe 23

Objectives 23
    Knowledge-Based 23
    Skill-Based 23
Keywords 23
Introduction 23
    Pipe Diameters 23
    Plastic Pipe 24
    PVC Pipe 24

  CPVC Pipe 24
  ABS Pipe 24
  Polyethylene Pipe 24
 Chapter Review Questions and Exercises 25
 Job Sheet 1: Pipe Types and Usage 27
  Instructor's Response 28
 Job Sheet 2: Common Lengths of Pipe 29
  Instructor's Response 30
 Job Sheet 3: Pipe Identification 31
  Instructor's Response 32

CHAPTER 4
Fittings 33
 Objectives 33
  Knowledge-Based 33
  Skill-Based 33
 Keywords 33
 Introduction 33
  Degree of Fittings 34
  Various Fitting Designs 34
 Chapter Review Questions and Exercises 35
 Job Sheet 1: Ordering a Plumbing Tee 37
  Instructor's Response 38
 Job Sheet 2: Creating a Reducing-Tee 39
  Instructor's Response 40

CHAPTER 5
Valves and Devices 41
 Objectives 41
  Knowledge-Based 41
 Keywords 41
 Introduction 41
  Isolation Valves 41
  Ball Valve 42

    Gate Valve. 42
    Stop and Waste 42
    Gas Cock 42
Chapter Review Questions and Exercises. 42
Job Sheet 1: Valve Identification. 45
    Instructor's Response 46
Job Sheet 2: Device Identification. 47
    Instructor's Response 48
Job Sheet 3: Isolation Valve 49
    Instructor's Response 50
Job Sheet 4: Stop and Waste Valve 51
    Instructor's Response 52
Job Sheet 5: Gas Cock 53
    Instructor's Response 54

CHAPTER 6
Fixtures 55

Objectives 55
    Knowledge-Based 55
    Skill-Based 55
Keywords. 55
Introduction. 55
    Toilets. 55
    Lavatory Sinks 56
    Bathtubs. 56
    Showers 56
Chapter Review Questions and Exercises. 56
Job Sheet 1: Lavatory Ordering Information 59
    Instructor's Response 60
Job Sheet 2: Bathtubs. 61
    Instructor's Response 62
Job Sheet 3: ADA Requirements for a Toilet. 63
    Instructor's Response 64
Job Sheet 4: Concrete Floor Penetrations 65
    Instructor's Response 66

CHAPTER 7
Faucets and Drain Assemblies 67

- Objectives 67
  - Knowledge-Based 67
  - Skill-Based 67
- Keywords 67
- Introduction 67
- Chapter Review Questions and Exercises 68
- Job Sheet 1: Faucet and Drain Questions 69
  - Instructor's Response 70
- Job Sheet 2: Faucet Identification 71
  - Instructor's Response 72
- Job Sheet 3: Bidet Drain Assembly 73
  - Instructor's Response 74

CHAPTER 8
Plumbing Equipment 75

- Objectives 75
  - Knowledge-Based 75
  - Skill-Based 75
- Keywords 75
- Introduction 75
  - Garbage Disposers 75
  - Dishwashers 76
  - Washing Machine Box 76
  - Icemaker Box 76
  - Residential Water Heaters 76
- Chapter Review Questions and Exercises 76
- Job Sheet 1: Garbage Disposal Identification 79
  - Instructor's Response 80
- Job Sheet 2: Tankless Water Heater 81
  - Instructor's Response 82
- Job Sheet 3: Determining if Food Waste Can Be Discharged into a Septic System 83
  - Instructor's Response 84
- Job Sheet 4: Determining the Dishwasher Drain Hose Requirements 85
  - Instructor's Response 86

CHAPTER 9
Blueprint Reading and Drafting 87

- Objectives 87
  - Knowledge-Based 87
  - Skill-Based 87
- Keywords 87
- Introduction 87
  - Drafting Tool 87
  - Scale Ruler 88
  - Drafting Triangles 88
  - Symbol Templates 88
  - Drafting Paper 88
  - Isometric Drafting 88
  - Riser Diagrams 88
- Chapter Review Questions and Exercises 89
- Job Sheet 1: Residential Plumbing Symbols 91
  - Instructor's Response 92
- Job Sheet 2: Using a Tape Measure for a Scale Ruler 93
  - Instructor's Response 94
- Job Sheet 3: Creating a Riser Diagram 95
  - Instructor's Response 96

CHAPTER 10
Material Organization and Layout 97

- Objectives 97
  - Knowledge-Based 97
  - Skill-Based 97
- Keywords 97
- Introduction 97
  - Communication 97
  - Written Communication 98
  - Oral Communication 98
  - Material Organization 98
- Chapter Review Questions and Exercises 99
- Job Sheet 1: Material Organization 101
  - Instructor's Response 102

Job Sheet 2: Concrete floor Penetrations 103
    Instructor's Response 104

CHAPTER 11
Water Service Installation 105

Objectives 105
    Knowledge-Based 105
Keywords 105
Introduction 105
    Water Source 105
    Public Water System 106
    Private Water Systems 106
    EPA Standards 106
    Water Quality 106
    Water Filtration 106
Chapter Review Questions and Exercises 107
Job Sheet 1: Water Meter Installation for Cold Weather Climates 109
    Instructor's Response 110
Job Sheet 2: Determining the Code Requirements for Working on a Public Water Supply ... 111
    Instructor's Response 112
Job Sheet 3: Water Service Installation Code Requirements 113
    Instructor's Response 114

CHAPTER 12
Water Distribution Installation 115

Objectives 115
    Knowledge-Based 115
Keywords 115
Introduction 115
    Layout and Sizing 115
    Pipe Sizing 116
    Sizing Theory 116
    Job Site Sizing 116
Chapter Review Questions and Exercises 116
Job Sheet 1: Brazing 119
    Instructor's Response 120

Job Sheet 2: Soldering 121
    Instructor's Response 122
Job Sheet 3: Flaring Copper Tubing 123
    Instructor's Response 124

CHAPTER 13
Drainage Waste, Vent Segments, and Sizing 125

Objectives 125
    Knowledge-Based 125
    Skill-Based 125
Keywords 125
Introduction 125
    Major Segments of a DWV System 126
    Minor Segments of a DWV System 126
Chapter Review Questions and Exercises 127
Job Sheet 1: Identify the Major Segments of a DWV System 129
    Instructor's Response 130
Job Sheet 2: Identify the Minor Segments of a DWV System 131
    Instructor's Response 132

CHAPTER 14
Drainage Waste and Vent Installation 133

Objectives 133
    Knowledge-Based 133
Keywords 133
Introduction 133
    Scope of Work 133
    Guest Bathroom Layout 133
    Kitchen Sink Layout 134
    Master Bathroom 134
    Hall Bathroom 134
    Laundry Room 134
    Building Drain 134
    Venting System 134
    Fixture Rough-in 134
    Toilets 134

Chapter Review Questions and Exercises.. 135
Job Sheet 1: Common DWV Layout Considerations 137
    Instructor's Response 138
Job Sheet 2: Wet Venting Code Requirements.. 139
    Instructor's Response 140

CHAPTER 15
Fixture and Equipment Installation 141

Objectives 141
    Knowledge-Based 141
Keywords... 141
Introduction... 141
    Escutcheons and Stops 141
    Toilets.. 141
    Lavatories... 141
    Kitchen Sinks 142
    Laundry Sinks 142
    Electric Water Heaters 142
    Gas Water Heaters 142
Chapter Review Questions and Exercises.. 142
Job Sheet 1: Common Tools and Items Required to Install a Toilet 143
    Instructor's Response 144
Job Sheet 2: Common Tools and Items Required to Install a Cultured Marble Lavatory ... 145
    Instructor's Response 146
Job Sheet 3: Common Tools and Items Required to Install a Compression Stop 147
    Instructor's Response 148

CHAPTER 16
Plumbing Repairs and Troubleshooting 149

Objectives 149
    Knowledge-Based 149
Keywords... 149
Introduction... 149
    Safety 149
    Electric Water Heaters 149
    High-Limit Devices 150

Upper Thermostat 150
Lower Thermostat 150
Heating Elements 150
Chapter Review Questions and Exercises 150
Job Sheet 1: Water Heater Identification 153
Instructor's Response 154
Job Sheet 2: Purging a System 155
Instructor's Response 156
Job Sheet 3: Electrical Wiring Requirements for a Plumber 157
Instructor's Response 158
Job Sheet 4: Disassemble of a Gas Regulator 159
Instructor's Response 160

CHAPTER 17
Hydronic Heat 161

Objectives 161
Knowledge-Based 161
Skill-Based 161
Keywords 161
Introduction 162
Theory of Hydronic Heating Systems 162
The Heat Source 162
Aquastat 162
Reset 162
Low Water Cutoff 162
Expansion Tanks 163
Centrifugal Pumps 163
Air Vent and Air Separators 163
Chapter Review Questions and Exercises 163
Job Sheet 1: Installing a Hydronic System 165
Instructor's Response 166
Job Sheet 2: Installing the Boiler 167
Instructor's Response 168

Job Sheet 3: Installing the Piping 169
    Instructor's Response 170
Job Sheet 4: Filling the System 171
    Instructor's Response 172

Preface

INTRODUCTION

Designed to accompany *Residential Construction Academy: Plumbing, Second Edition*, this workbook is an extension of *Plumbing, Second Edition* core text by providing additional review questions and exercises designed to challenge and reinforce the student's comprehension of the content presented in the core text.

ABOUT THE TEXT

The Workbook is divided into chapters, with each chapter directly corresponding to a chapter in *Residential Construction Academy: Plumbing, Second Edition*. Each chapter consists of an introduction, objectives, review questions, job sheets, and exercises. The review questions are composed of a variety of matching, true false, multiple choice, short answer, and essay-style questions based on the materials presented in the core text and workbook.

JOB SHEETS

Each job sheet consists of an objective for that job sheet, instructions, and either an activity or a checklist. The job sheets range in complexity from entry level to more complex problems that require the student to perform calculations.

FEATURES OF THIS WORKBOOK

- Additional review questions and exercises for *Residential Construction Academy: Plumbing, Second Edition*
- Job sheets with additional exercises and activities designed to reinforce the material presented in the core text book

Plumber's Toolbox

CHAPTER 1

OBJECTIVES
Upon completion of this chapter, you will be able to:

Knowledge-Based
- Identify and describe the hand tools the plumber commonly uses.

Skill-Based
- Use hand tools in a safe and appropriate manner.

KEYWORDS

band iron
cheater
fitting
joint
nut driver
open-end wrench
slope
socket
wallboard
wire cutter

INTRODUCTION

Some tools and equipment are common to all plumbing trades (residential, commercial, or industrial). Therefore, understanding their intended usage is essential to safely completing a particular task. In addition, it is imperative that plumbing students understand how to properly care for their tools. Keeping plumbing tools in good operating condition will not only make completing a plumbing task easier, but it will also help prevent workplace accidents. Faulty tools are one of the leading causes for injuries in the building trades industry.

HAND TOOLS

Hand tools, such as hammers, screwdrivers, levels, and tapes, do not use a motor. They are usually less expensive compared with power tools but are just as important. A few of these hand tools are outlined below.

LEVELS

- A level is a critical tool in any plumber's toolbox; in fact, it is one of the most important tools.
- Levels typically contain a vial filled with a colored liquid and a trapped air bubble.

- A moat level has dimensions marked along at least one of its edges and can be used for measuring distances.
- Two types of levels most commonly used by plumbers are the torpedo level and the 24" level.
- Most plumbing levels are magnetized on one side for installing metal pipe.

MEASURING

Accuracy and care in measuring are all-important. They can mean the difference between a well-put-together project and a sloppy one.

TAPE MEASURES

- Tape measures are available in a variety of different lengths.
- The most common length of tape measure used by plumbers is the 25-foot retractable tape measure.
- The wider the blade of a tape measure the greater the distance the plumber can measure without the tape measure collapsing.

SQUARES

- Framing squares are used extensively by both plumbers and carpenters for layout work.
- They consist of two sides that form a 90° angle.
- Speed squares are a common tool for marking boards to cut.
- Both types of squares have dimensions for performing measurements and layouts.

SCREWDRIVERS

- They are available in a variety of different styles, lengths, and shank diameters.
- Screwdrivers most commonly used by plumbers are the Phillips and the slotted-head.
- The tip of the screwdriver determines the size.
- The most common sizes used are #1, #2, and #3.
- The most common type of screwdriver is the multi-screwdriver.
- The multi-screwdriver can be disassembled and reassembled for specific uses.

PLIERS

- Pliers are one of the most practical tools that plumbers can have in their toolbox.
- The most common type of pliers used by plumbers is the angled and grooved jaw.
- The 10" angled-jaw pliers can be adjusted to an opening of up to 2".
- Angled pliers are available in a variety of different sizes.
- Combination pliers are handy for working in small tight places.
- Locking pliers are handy for removing nuts and bolts that have seized due to corrosion.
- Needle-nose pliers are useful for working with small parts.

ADJUSTABLE WRENCHES

- Adjustable wrenches are available in a variety of different sizes.
- The jaws of the adjustable wrench are smooth so it will not damage metal finishes.
- A common size of adjustable wrench used by plumbers is the 9".

CHAPTER REVIEW QUESTIONS AND EXERCISES

TRUE/FALSE

1. _____ The right tool can help a plumber complete a task more efficiently and safely. However, when the correct tool is not available, it is perfectly acceptable to use a tool for something other than what it was designed for as long as the plumber uses extreme caution.

2. _____ When working with chemicals or solvents, all of the safety precautions and data included on product labels must be read and appropriate inhalation protection equipment used.

3. _____ The plumbing trade uses only specialty tools.

4. _____ The tools used in the residential industry are different from the tools used for commercial plumbing; therefore, if a plumber wishes to do both residential and commercial plumbing, two totally different sets of tools will be required.

5. _____ Mesh glove liners can be purchased and worn under protective gloves to provide warmth in cold weather.

6. _____ A torque wrench is used to tighten clamps used for installing cast iron pipe as well as for tightening rubber transition connectors used for dissimilar piping.

7. _____ Safety on the job site is the responsibility of only that trade's supervisor.

8. _____ A listing of all safety hazards and medical attention necessary for a specific product can be found on the Material Safety Data Sheet (MSDS).

9. _____ An MSDS must be available for all hazardous products and kept on file at the job site.

10. _____ Any alteration of a hard hat including but not limited to applying logos and stickers is an OSHA violation.

COMPLETION

11. Which personal protection equipment (PPE) is mostly relevant to hand tools?

_____

_____

_____

12. What are flaring tools used for?

13. As mentioned in the textbook, which OSHA standard regulates the construction industry?

14. Why is it important to use the correct tool for a specific task?

15. What is the most frequently found PPE on a residential job site?

16. Many styles of goggles have ventilation holes located in the top of the frame. What is the purpose of these ventilation holes?

Name: _____ Date: _____

Job Sheet 1: The Plumber's Toolbox

- Upon completion of this job sheet, you should be able to identify the tools commonly used in all plumbing trades.
- This job sheet is a checklist of the basic tools used by all plumbing trades.

Tool	Present	Condition
24" toolbox	☐	_____
24" pipe wrench	☐	_____
18" pipe wrench	☐	_____
12" angled-jaw pliers	☐	_____
Large slotted screwdriver	☐	_____
Medium slotted screwdriver	☐	_____
Medium Phillips screwdriver	☐	_____
Multi-screwdriver	☐	_____
5/16" nut driver and/or no-hub torque wrench	☐	_____
24" pipe level	☐	_____
Torpedo pipe level	☐	_____
Claw hammer	☐	_____
Straight-cut aviation snips or a set of three	☐	_____
25'–0" retractable tape measure	☐	_____
6'–0" folding rule	☐	_____
8" or 10" adjustable wrench	☐	_____
Hacksaw	☐	_____
Plumb bob	☐	_____
Chalk box	☐	_____
Concrete chisel	☐	_____
Wood chisel	☐	_____
Allen wrench set	☐	_____
Copper midget tubing cutter	☐	_____
Copper tubing cutter from 1/8" to 1⅛"	☐	_____
Copper tubing cutter up to 2" pipe size	☐	_____
Utility knife	☐	_____
Small soldering torch regulator kit	☐	_____
Torch striker	☐	_____
B-tank soldering torch kit	☐	_____

INSTRUCTOR'S RESPONSE

Name: _____ Date: _____

Job Sheet 2: The Plumber's Toolbox

- Upon completion of this job sheet, you should be able to identify the specialty tools used in all plumbing trades.
- This job sheet is a checklist of the basic tools used by all plumbing trades.

Tool	Present	Condition
Pipe nipple extracting set	☐	_____
Thread tapping tool	☐	_____
Inside plastic pipe cutter	☐	_____
Basin wrench	☐	_____
Mini-hacksaw	☐	_____
Nail-puller	☐	_____
Smooth-jaw pipe wrench	☐	_____
Plastic pipe saw	☐	_____
Needle-nose pliers	☐	_____
Locking pliers	☐	_____
6" combination pliers	☐	_____
Copper flaring tool	☐	_____
Copper tubing bender	☐	_____
Flexible tubing cutter	☐	_____
Copper swaging tool	☐	_____
Basket strainer tool or internal wrench	☐	_____
Multi-purpose knife/pliers tool	☐	_____
Carpenter square	☐	_____
Metal stud punch	☐	_____
Strap pipe wrench	☐	_____
Chain pipe wrench	☐	_____
Miniature hacksaw	☐	_____
Wallboard saw	☐	_____
Flexible tubing crimping tool	☐	_____
Ball-peen hammer	☐	_____
Inside plastic pipe cutter	☐	_____
Cast iron chain cutters	☐	_____
Internal cast iron cutters	☐	_____

INSTRUCTOR'S RESPONSE

Name: _____   Date: _____

Job Sheet 3: Hand Tools

- Upon completion of this job sheet, you should be able to demonstrate your ability to safely use hand tools.

1. Compare four different brands of tape measures.
   a. How are the symbols different from one tape measure to another and what do the symbols mean?

   b. What are the increments of the scales?

2. Check your combination square for accuracy.
   a. Tape a sheet of paper to a board that has a perfectly straight edge.
   b. Hold the square against the edge and draw a line along the outer edge of the blade.
   c. Flip the square over so the opposite side of the blade faces up, align the square on the edge of the stock, and draw a second line about 1/32" from the first.
   d. If the square is accurate, the lines will be parallel.

3. Investigate the different types of hand saws available at your local home improvement center and write a brief description of the physical differences of the saws and how each one should be used.

INSTRUCTOR'S RESPONSE

Name: _____ Date: _____

Job Sheet 4: Using a Level to Set the Slope of a Pipe

- Upon completion of this job sheet, you should be able to demonstrate your ability to use a level in setting the slope of a pipe.

The textbook mentions using a 1/2" thick piece of wood on a 24" level to set the slope of a pipe to 1/4" per foot. Refer to this section of the textbook to answer the following questions:

1. Will the same thickness of wood work for a slope greater than 1/4" per foot? (Explain your answer.)

2. Can a shorter level be used to achieve the same effect? (Explain your answer.)

INSTRUCTOR'S RESPONSE

Name: _____   Date: _____

Job Sheet 5: First Aid Kits

- Upon completion of this job sheet, you should be able to identify the items necessary so that a first aid kit is compliant with ANSI Z308.1 standards.

Using the Internet and your textbook (Chapter 1, "First Aid Kit") list the items necessary for a first aid kit to be considered ANSI Z308.1 compliant.

INSTRUCTOR'S RESPONSE

Power Tools

CHAPTER 2

OBJECTIVES

Upon completion of this chapter, you will be able to:

Knowledge-Based
- Identify and describe power tools often used by a plumber.
- Identify which drill or saw is used relevant to the work location.
- Identify which drill bit or saw blade is used relevant to a system installation.
- Know the proper personal protection equipment (PPE) required for power tools.

Skill-Based
- Use power tools in a safe and appropriate manner.

KEYWORDS

Allen screw
bit
chuck
chuck key

flashing
floor joist
lanyard
OSHA

shank
wall stud

INTRODUCTION

In addition to working with hand tools, a plumber will be required to work with power tools and in potentially dangerous conditions. Knowing how to properly care for and use power tools is essential if the plumber is to complete an assigned task without injuring himself or herself or others. In addition, situations occasionally arise in which the plumber will be required to work in unfriendly environmental conditions. Therefore, it is imperative that the plumbing technician has a solid understanding of working safely.

ELECTRICAL SAFETY

- Because electrical power tools are connected directly to an electrical power source, you should never use them in damp or wet conditions.
- OSHA requires all power tools to be connected to a ground fault circuit interrupter (GFCI).
- If a power tool should get wet, then have it inspected by a qualified service technician before attempting to use it.
- It is critical that all GFCIs be tested using a portable GFCI test on a regular basis to ensure they are operating correctly.

LADDER SAFETY

- When a plumber is working from a ladder the possibility that he or she could incur a serious injury or even death increases.
- The possibility of injury or death increases even more when power tools are used while working from a ladder.
- When using a step ladder, never stand on the top step; doing so is a violation of OSHA regulations.
- Never exceed the maximum weight limit of a ladder; therefore always consider the weight of any materials and tools that will be used while on the ladder as well as your own weight before using the ladder.
- The extension ladder must extend 3 feet above the point at which they contact the edge of the roof.
- The bottom of an extension ladder should be placed out from its point of contact by a distance of approximately one-quarter of the length of the ladder.

FALL PROTECTION

- According to OSHA, fall protection is determined by the location and type of work being performed.
- Fall protection must be employed whenever your feet are 6 feet or more above the ground.
- The most common method of fall protection used is the body harness and a securing lanyard.

EAR PROTECTION

- Often damage to the inner ear is not apparent until several years after exposure to loud noises.
- Ear protection is one of the PPE items that will protect a plumber from long-term injury.
- If you have a ringing sensation after wearing hearing protection and operating powers tools, then it is an indication that your hearing protection is inadequate.

DRILLS

- A plumber uses drills to bore holes in various building materials.
- Three drills often used by plumbers to drill wood products are the pistol, the right angle, and the hole hawg.
- Never use a drill when wearing loose clothing.
- If you are wearing long sleeves while operating a drill, always keep your sleeves buttoned.

RIGHT ANGLE DRILL

- The right angle drill is perfect for drilling holes between the floor joist and/or wall studs.
- The chuck of a right angle drill is positioned 90° to the body of the drill.

PISTOL DRILL

- A pistol drill is so named because its design makes it look like a handgun.

HOLE HAWG

- This is another type of right angle drill that has a fixed drill chuck positioned further in the drill's body, making it more compact.

CHAPTER REVIEW QUESTIONS AND EXERCISES

TRUE/FALSE

1. __T__ Always follow the manufacturer's instructions for use and maintenance of all electrical tools and appliances.
2. __F__ Ladders shall be securely fixed at the top and foot so that they cannot move either from their top or from their bottom points of rest.
3. __F__ If it is not possible to secure a ladder at both the top and bottom, then it shall be securely fixed at the top.
4. __T__ Power tools must be used by trained personnel only.
5. __F__ A grinder rotates at high speed and cannot grab loose clothing. Therefore, it is not important to button long-sleeved shirts before using a grinder.
6. __F__ Taping a torpedo level to a drill with duct tape will ensure that the hole is drilled correctly.
7. __F__ A saber saw is primarily used by plumbers to cut holes in countertops to install sinks.

COMPLETION

8. What is the primary purpose of a portable band saw?

    CUT ALL THREAD FOR HANGERS

Name: _____ Date: _____

Job Sheet 1: Power Tool Checklist

- Upon completion of this job sheet, you should be able to identify the power tools commonly used in the plumbing trades.

Tool	Present	Condition
Circular saw	☐	_____
Portable band saw	☐	_____
Saber saw	☐	_____
Right angle drill	☐	_____
Jackhammer	☐	_____
Air compressor	☐	_____
Powder-actuated tool	☐	_____
Grinder	☐	_____
Hole hawg	☐	_____
Pistol drill	☐	_____
Reciprocating saw	☐	_____
Hammer drill	☐	_____

INSTRUCTOR'S RESPONSE

Name: _____ Date: _____

Job Sheet 2: Right Angle and Hole Hawg Safety Instructions

- Upon completion of this job sheet, you should be able to understand the manufacturer's safety instructions for right angle and hole hawg drills.

PROCEDURE

Compare the manufacturer's safety instructions for a right angle drill to the manufacturer's safety instructions for a hole hawg drill.

1. What part(s) of the safety instructions are the same?

   _____

   _____

   _____

   _____

2. How do the safety instructions differ?

   _____

   _____

   _____

   _____

3. What is the manufacturer's recommendation in case the tool gets wet?
   a. For the right angle drill

   _____

   _____

   _____

   _____

   b. For the hole hawg drill

   _____

   _____

   _____

   _____

INSTRUCTOR'S RESPONSE

Types of Pipe

CHAPTER 3

OBJECTIVES
Upon completion of this chapter, you will be able to:

Knowledge-Based
- Identify and describe common types of pipe and tubing used in a residential plumbing installation.
- Understand that certain pipe and tubing materials can be used only for specific systems.
- Understand that some pipe and tubing materials can be used in all residential systems.
- Relate pipe and tubing selection to plumbing codes.

Skill-Based
- Order pipe and tubing materials.

KEYWORDS

drain
fitting
foam-core
nominal diameter
offset
pipe
potable water
radius
schedule
sewer
tube
tubing
type
vent
water distribution
water service

Introduction

Regardless of the type of work a plumber is asked to do, a good understanding of the products and materials available is essential to effectively design a plumbing system that meets the client's (property owner's) needs. This includes the types of piping available and the local building codes that regulate their usage. Pipe is used to bring potable water and gas into a residence and to allow sewage and wastewater to drain from a residence.

Pipe Diameters

- In plumbing, piping is specified by its diameter.
- Piping and tubing are ordered using a system known as nominal diameter.
- Both the inside diameter (ID) and outside diameter (OD) will vary based on the material from which the piping or tubing is made.

- When determining flow requirements the ID is an important factor; in some cases the plumber orders pipe based on its nominal ID.
- Copper tubing is used in both the HVAC and plumbing industries.
- Plumbers order copper tubing using both its ID and OD while HVAC technicians order copper tubing based on its OD.

Plastic Pipe

- The introduction of pipe as an alternative for drainage, vent, and water distribution has reduced the cost of residential construction.
- Plastic pipe is typically easier to install and therefore has reduced the cost of labor.
- Plastic pipe is not fire resistant.
- Most residential homes use a combination of metal and plastic products.

PVC PIPE

- Polyvinyl chloride (PVC) pipe is available in a variety of types and schedules.
- PVC pipe used in residential construction has a maximum temperature rating of 140°F.
- Pipe and tubing used in residential water service must be able to handle a pressure of 160 psi and a temperature of 73°F.
- Pipe and tubing used for water distribution systems must be able to handle a pressure of 100 psi at 180°F.

CPVC PIPE

- Pipe made of chlorinated polyvinyl chloride (CPVC) is flexible, but is typically called pipe, not tubing.
- CPVC is a yellowish-white material that is joined using a solvent weld process (glue). Because PVC and CPVC are incompatible, they require different glues.
- The OD of CPVC pipe is not the same as that of PVC pipe, and their nominal sizes do not reflect their actual sizes.
- CPVC used in residential construction ranges in size from 1/2" to 2".
- Long-term exposure to UV rays can damage CPVC; therefore CPVC pipe must be covered (or painted).

ABS PIPE

- Acrylonitrile butadiene styrene (ABS) pipe is black in color and is used to install drainage waste and vent (DWV) systems.
- ABS pipe is not compatible with PVC pipe.
- ABS pipes must be joined with a solvent weld process designed specifically for that product.
- ABS pipes range in size from 1-1/2" to 6".
- ABS pipes have a maximum water-temperature capability of 140°F.

POLYETHYLENE PIPE

- Polyethylene (PE) tubing is used for water service installations but is not allowed by code to be installed on the interior of a residential building.
- It is usually sold in 100' rolls.

CHAPTER REVIEW QUESTIONS AND EXERCISES

SHORT ANSWER

1. How are copper connections soldered?

TRUE/FALSE

2. _____ Brass pipe and fittings can be used for all piping systems.

3. _____ PVC should never be used for drainage waste and vent (DWV) applications. Explain your answer.

4. _____ Pipes are manufactured to specific standards. Explain your answer.

5. _____ PVC cannot be used for water distribution inside a residential structure. Explain your answer.

COMPLETION

6. _____ and _____ are the organizations that rate materials.

7. Cross-linked polyethylene is a flexible tubing used for the installation of _____.

8. The brazing of copper connections is known as _____.

9. _____ and _____ types of pipe are used for gas piping systems.

10. What are the two types of cast iron pipes used for DWV systems?

Name: _____ Date: _____

Job Sheet 1: Pipe Types and Usage

- Upon completion of this job sheet, you should be able to identify various pipe types and their usage.

Using the following chart, identify the various types of pipes and their usage.

Material Type	DWV	Potable Water
PVC		
ABS		
CPVC		
PEX		
Copper		
Galvanized steel		
Black steel		
Brass		
Cast iron		
PE		
Perforated PVC		

INSTRUCTOR'S RESPONSE

Name: _____   Date: _____

Job Sheet 2: Common Lengths of Pipe

- Upon completion of this job sheet, you should be able to identify common lengths of pipe sold.

Using the following chart and information from a plumbing distributor or home improvement store, identify the various lengths in which each pipe can be obtained.

Note: National plumbing distributors such as a home improvement store may be contacted through their websites.

Material Type	30"	5'-0"	10'-0"	20'-0"	21'-0"	Roll
PVC						
ABS						
CPVC						
PEX						
Copper						
Galvanized steel						
Black steel						
Service-weight cast iron						
No-hub cast iron						
PE						
Perforated PVC						
Brass						

Distributor ____ _____

INSTRUCTOR'S RESPONSE

Name: _____   Date: _____

Job Sheet 3: Pipe Identification

- Upon completion of this job sheet, you should be able to identify the correct piping material for a particular task.

Match the pipe material to the application.

A. Copper	_____	1.	_____ pipe cannot be used for potable water distribution. It is used for residential gas supply piping.
B. Cast iron	_____	2.	_____ pipe is used for water service installations but is not allowed by code to be installed on the interior of a residential building.
C. Black steel	_____	3.	A _____ is used for water service installations but is not allowed by code to be installed on the interior of a residential building.
D. Galvanized steel	_____	4.	A _____ system must be tested with water, never air.
E. PVC	_____	5.	_____ pipe is sold in hard and soft versions and used in most plumbing applications. In addition, it is used by the HVAC industry.
F. ABS	_____	6.	_____ pipe is often used in residential DWV systems for vertical installations to allow a quieter draining process in walls.
G. PE Pipe	_____	7.	_____ pipe is offered in a wide variety of types and schedules and is the most widely used product for DWV installations in new residential applications.
H. PEX	_____	8.	_____ pipe can also be used for DWV systems and was the material of choice in the early and middle twentieth century.
I. CPVC	_____	9.	_____ pipe is used for water distribution systems and has become one of the most popular selections in the housing industry because of its cost and labor efficiency.

INSTRUCTOR'S RESPONSE

CHAPTER 4

Fittings

OBJECTIVES

Upon completion of this chapter, you will be able to:

Knowledge-Based
- Identify and describe common types of fittings in a residential plumbing installation.
- Understand that certain fitting materials and designs can be used only for specific systems.
- Understand that some fitting materials can be used in all residential systems.
- Relate fitting design selection to plumbing codes.

Skill-Based
- Order fittings based on installation requirements, size, and materials.

KEYWORDS

adapter	electrolysis	stack
bell	female	street
bend	hub	swing joint
bushing	joint	tee
cleanout	male	test tee
closet bend	offset	trap adapter
combo	p-trap	union
coupling	reducer	vent
drain	sanitary cross	water closet
DWV	sanitary tee	wye
elbow	socket	

INTRODUCTION

Regardless of the type of piping materials used in a plumbing system, the basic components (fittings) used to connect the piping together into a useable system are the same (just made from different materials). The fittings in a piping system provide creative installation routes. Offsets and termination connections within piping systems are completed with fittings that are designed for specific uses. Some fitting materials must be made of the same material as the pipe, while other approved materials are interchangeable with numerous systems and types of pipe. Listed below are some of the more common fittings used in residential plumbing.

- 90° elbow—used to change the direction of a pipe by 90°.
- 45° elbow—used to change the direction of a pipe by 45°. It is often used in pairs to offset a section of pipe around an obstacle.

- Tee—used to create a branch line.
- Cap—used to terminate a section of pipe. A cap is a fitting that fits around the outside of a pipe.
- Coupling—used to connect two pieces of pipe having the same diameter.
- Plug—used to terminate a section of pipe. A plug is a fitting that screws into the inside of a section of pipe.
- Union—used to connect two pieces of pipe together, while allowing for the two pieces of pipe to be disconnected without having to cut the pipe into two sections.
- Reducers—fitting used to connect two different pipe sizes together. Reducers are different than bushings but accomplish the goal of reducing a pipe size.
- Bushing—used to reduce a section of pipe similar to a reducer, but threads inside a fitting to create the reduction.

Degree of Fittings

- The three most common offsets used by plumbers today are the 22-1/2°, the 45°, and the 90°.
- The 22-1/2° is used in drainage waste and vent (DWV) systems.
- 45° fittings are not used in a flexible tubing system because the tubing can be bent to accommodate these offsets.
- A cast iron 90° fitting is ordered as a quarter bend.
- A cast iron 45° fitting is ordered as an eighth bend.
- A cast iron 22-1/2° fitting is ordered as a sixteenth bend.

Various Fitting Designs

Before installing a piping system the plumber must have a good working knowledge of the following:
- Fittings, building codes, job-site conditions, and the correct material to use for a specific system.
- Fittings are named based on their unique characteristics and material type.
- Fittings are classified based on their design.

There are two basic connection types for fittings:
- Fittings that can receive a pipe (hub) and fittings that can be inserted into a hub (socket) of a pipe end.
- No-hub cast iron (NHCI) fittings are connected using a specifically designed clamp.

OFFSETS

- The purpose of an offset is to change the direction of a piping system.
- Pressure systems use shorter (tighter) radius changes while DWV systems use larger radius.
- Two 45° fittings can be used to create a 90° offset.
- In a DWV system, two 22-1/2° fittings can be used to create a 45° offset.
- When an offset is created using two fittings it is referred to as a "swing joint."

TEES

- A fitting with three connections used for pressure systems is known as a tee.
- Pressure systems use a different flow pattern and a DWV system.
- For sizing and ordering purposes the opening of a tee is identified as either run or branch.
- Run openings are those in the direction of flow through a tee.
- The branch opening is perpendicular to the run opening.

CHAPTER REVIEW QUESTIONS AND EXERCISES

TRUE/FALSE

1. Most fittings are manufactured from the same material as the connecting pipe. (Explain your answer.)

2. Brass fittings are not interchangeable with numerous piping materials. (Explain your answer.)

3. Threaded fittings cannot be used to connect different material. (Explain your answer.)

4. Plumbing codes dictate that solvent-welded fittings can be used only with compatible pipe. (Explain your answer.)

5. Dissimilar metal connections can cause corrosion and must be protected against electrolysis. (Explain your answer.)

_____

_____

_____

_____

6. DWV fittings have the same flow pattern as pressure fittings. (Explain your answer.)

_____

_____

_____

_____

COMPLETION

7. A(n) _____ is a vertical drain or vent pipe rising more than one story high.

8. _____ is another name for a toilet.

9. A(n) _____ is a fitting arrangement to create an offset using two fittings.

10. _____ is a corrosion process caused from directly connecting dissimilar metals.

11. A(n) _____ is an enlarged end of a pipe or fitting that receives a pipe end or fitting and that may also be called a bell or a hub.

12. A pipe that receives discharge from a fixture(s) is a _____.

13. A(n) _____ is a fitting having internal threads and screws over a male fitting.

14. A type of offset fitting that has one end with the same outside diameter as a connecting pipe or fitting is called a _____.

15. A(n) _____ is a pipe dedicated to providing airflow so that a drainage system can breathe.

Name: _____ Date: _____

Job Sheet 1: Ordering a Plumbing Tee

- Upon completion of this job sheet, you should be able to order a plumbing tee.

For each tee, provide the correct ordering information.

Side 1	Side 2	Side 3	Order as
1/2"	1/2"	1/2"	
1/2"	1/2"	3/4"	
3/4"	3/4"	3/4"	
3/4"	3/4"	1/2"	
3/4"	1/2"	1/2"	
3/4"	1/2"	3/4"	
3/4"	3/4"	1"	
1"	1"	1"	
1"	1"	3/4"	
1"	1"	1/2"	
1"	3/4"	3/4"	
1"	3/4"	1/2"	
1"	1/2"	1/2"	

INSTRUCTOR'S RESPONSE

Name: _____   Date: _____

Job Sheet 2: Creating a Reducing Tee

- Upon completion of this job sheet, you should be able to create a reducing tee (if a reducing tee is not available).

PROCEDURE

1. List the steps for creating a reducing tee.

2. Using the materials supplied by your instructor create a reducing tee.

INSTRUCTOR'S RESPONSE

Valves and Devices

CHAPTER 5

OBJECTIVES
Upon completion of this chapter, you will be able to:

Knowledge-Based
- Identify and describe valves and devices used in a residential plumbing installation.
- Understand that certain valve designs can be used only for specific systems.
- Know the safety devices used in residential piping systems and their unique characteristics.
- Relate installation of valves and devices to plumbing codes.

KEYWORDS

air gap	gas cock	reactionary valves
anti-siphon	gate valve	reduced-pressure zone valve
backflow	hose bibb	relief valve
back siphon	hose outlet	stop
ball valve	isolate	stop-and-waste valve
boiler drain	isolation valve	T&P valve
check valve	lug	vacuum breaker
flush	pressure-reducing valve	vacuum-relief valve

INTRODUCTION

Regardless of the application, all plumbing systems use devices and valves. Valves are used to control the flow of water, waste, gas, and so on. They can be operated manually or automatically by using motors and a computer interface to control their actions. Valves and devices installed for potable water must be approved by plumbing codes. Threaded valves and devices typically have female threads while soldered connections are used for many valves and devices connecting to copper tubes. Plastic valves and devices are available with solvent welded connections. A valve design used to isolate a single fixture is known as a stop, while a valve design specifically used to isolate gas supply to a fixture is known as a gas cock. Backflow devices are installed to protect a potable water system; a pressure-reducing valve reduces the pressure in a piping system.

ISOLATION VALVES

- The code requires every residence to have at least one isolation valve.
- Isolation valves must be installed in a location in which the water can be cut off in case of an emergency or a repair.

- Most codes dictate that the minimum size of a residential water service is 3/4".
- Many plumbing codes require that the main isolation valve for a residence be a full port design.
- A full port valve has the same inside diameter as the pipe connecting to it.
- Most devices and some valves have a direction of flow and therefore require that the plumber know the flow direction of the water or gas.

Common Isolation Valves Used in Residential

Type	Residential Uses
Ball valve	Water and gas
Gate valve	Water
Stop valve	Water
Stop-and-waste valve	Water
Gas cock	Gas

BALL VALVE

- A ball valve utilizes an internal ball with a hole in its center.
  - It creates a flow passageway through the valve.
  - It isolates flow when the ball is rotated 90° from the flow direction.

GATE VALVE

- A gate valve utilizes a metal gate (disc) that slides vertically to open and close the valve.

STOP-AND-WASTE VALVE

- A stop-and-waste valve uses the same design as a stop valve to isolate an entire water distribution system, except that it also has a draining feature.

GAS COCK

- A gas cock is used for gas distribution systems.

CHAPTER REVIEW QUESTIONS AND EXERCISES

COMPLETION

1. The dangerous reversal of flow in a piping system, which can contaminate that system, is called _____.

2. A(n) _____ is the smooth, internal surface of a valve or device that, when mated with a washer or another isolating feature, stops flow in a piping system.

3. A device that prevents siphoning of contaminates into a piping system is a(n) _____.

4. _____ is separating a portion of a piping system by turning a valve or device.

5. A(n) _____ is a type of vacuum breaker that is commonly used on a water heater that is piped with a side inlet connection.

6. Unobstructed vertical space from a device outlet to a point where water could backflow into a piping system is called _____.

7. _____ is cleaning a piping system with air or water pressure.

8. A designated raised portion of a valve and device used in place of a handle to operate with a wrench or tool is a(n) _____.

Name: _____ Date: _____

Job Sheet 1: Valve Identification

- Upon completion of this job sheet, you should be able to correctly identify various valves used in plumbing.

For each valve, give a brief description of what that valve is used for.

Type	Residential Uses
Ball valve	
Gate valve	
Stop valve	
Stop-and-waste valve	
Gas cock	

INSTRUCTOR'S RESPONSE

Name: _____ Date: _____

Job Sheet 2: Device Identification

- Upon completion of this job sheet, you should be able to correctly identify various devices used in plumbing.

For each valve, give a brief description of what that device is used for.

Type	Use
Pressure-reducing valve	
Vacuum breaker	
Vacuum-relief valve	
Relief valve	
Reduced-pressure zone valve	
Double-check valve assembly	

INSTRUCTOR'S RESPONSE

Name: _____   Date: _____

Job Sheet 3: Isolation Valve

- Upon completion of this job sheet, you should be able to demonstrate your understanding of isolation valves.

ISOLATION VALVES

1. True or False: Every residential dwelling is required by code to be provided with at least two isolation valves.

2. True or False: The valve must be installed in a readily accessible location so the homeowner can shut off the water supply in case of an emergency or a repair.

3. What is the minimum size dictated by most building codes for a residential water service?

4. True or False: Some isolation valves and most devices installed in a piping system have a direction of flow and require an installer to connect the piping system knowing the flow direction of the water or gas.

INSTRUCTOR'S RESPONSE

Name: _____ Date: _____

Job Sheet 4: Stop-and-Waste Valve

- Upon completion of this job sheet, you should be able to demonstrate your understanding of stop-and-waste valves.

    1. True or False: A stop-and-waste valve is an isolation valve that also has the capability to manually drain the isolated portion of a system.

    2. Many kinds of valves can be installed below ground in specially designed valve boxes; it is illegal and a violation of all codes to install a stop-and-waste valve below ground.

    3. True or False: A stop-and-waste valve should not be installed where a backflow of nonpotable water could enter the water distribution system through the drain port while the system is not under pressure.

INSTRUCTOR'S RESPONSE

Name: _____   Date: _____

Job Sheet 5: Gas Cock

- Upon completion of this job sheet, you should be able to demonstrate your understanding of gas cocks.

1. What is the main purpose of a gas cock?

2. True or False: Many gas cock designs do not have a manual handle such as a lever or wheel handle, but instead require a wrench to open and close the gas cock.

3. True or False: Most system isolation valves are located in the interior of a building near a gas meter.

4. The _____" and _____" sizes are the most commonly used gas cocks for residential applications and have female threaded connections.

5. True or False: Most gas cocks used in conjunction with a gas meter have an alignment hole in which to place a padlock to secure the gas distribution system when not in use.

INSTRUCTOR'S RESPONSE

Fixtures

CHAPTER 6

OBJECTIVES
Upon completion of this chapter, you will be able to:

Knowledge-Based
- Identify the basic types of residential fixtures.
- Understand that some fixtures are installed at different phases of construction.
- Recognize the importance of manufacturer installation information.

Skill-Based
- Order each fixture based on type and variations.

KEYWORDS

custom
pop-up
rough

rough-in
rough-in sheet
stub out

submittal
trim out

INTRODUCTION

Calculating and installing pipe is only a small portion of a plumber's job; plumbers must also be concerned with plumbing fixtures. A plumbing fixture by definition is a device that is part of a system to deliver and drain away water, but that is also configured to enable a particular use. Plumbing codes state that every home must have a minimum of at least one toilet, lavatory sink, bathtub or shower, and a kitchen sink. Plumbing codes also regulate the materials used to manufacture plumbing fixtures, which must have smooth, impervious surfaces and be defect free. Also clearances from walls and other fixtures are regulated by plumbing code.

TOILETS

- Toilets are available in a variety of styles and colors; however, the selection of a toilet is based on the cost of the home as well as the homeowner's preference.
- Two-piece combination toilets consist of a bowl and a tank.
- One-piece toilets are used in more expensive homes and often are used in only one bathroom, typically the master bedroom.

- To adhere with water conservation regulations, a toilet must use a maximum of 1.6 gallons per flush (gpf).
- Another water-saving feature that may be present on a toilet is the two flushing mode feature.
- For liquid waste there is an option of 0.08 gpf (half-flush).
- For solid waste there is an option of 1.6 gpf (full-flush).
- The siphon-jet flushing action is the most common design used for residential toilets.

LAVATORY SINKS

- A sink located in a bathroom is referred to as a lavatory and is often called a basin.
- Lavatory sinks are available in a variety of different styles, colors, and types.
- Lavatory sinks are ordered based on their size, color, shape, and mounting requirements.
- A drop-in sink requires a hole to be cut into the countertop and the sink is then installed into the hole.
- An under-counter sink still requires a hole to be cut into the countertop but the sink is installed from the underside of the countertop.

BATHTUBS

- A standard residential tub is 5' long and 30" wide.
- The water depth a tub can hold will vary based on the design of the tub.
- The capacity of a tub is determined by multiplying the length by width by depth.
- The volume of a tub determines how much water is needed to fill the tub.
- The tub overflow determines the actual depth of the water in the tub.

SHOWERS

- Codes dictate that a non-handicap shower must be at least 30" × 30" and have a threshold to prevent water from spilling over onto the area outside the shower.
- Showers are sold in a variety of styles and colors.

CHAPTER REVIEW QUESTIONS AND EXERCISES

TRUE/FALSE

1. _____ The rough-in phase of construction is before the wall finishes are installed.

2. _____ The trim-out phase of construction is the first installation phase.

3. _____ Most fixtures are provided with a manufacturer rough-in sheet and installation instructions.

4. _____ A tub and shower unit is a two-piece fixture.

5. _____ Most bathtubs are installed before the wall finish is complete.

6. _____ A drop-in-type tub is typically installed on a platform that is constructed by a carpenter.

7. _____ A kitchen sink is installed during the trim-out phase of construction.

8. _____ The capacity that some fixtures hold is not a determining factor in calculating gallons of water used.

9. _____ Bidets are available in styles that match an adjacent toilet.

10. _____ Laundry sinks are typically installed in the same room as a washing machine.

11. _____ Most kitchen sinks are offered as a four-hole design, but, if a separate handheld sprayer is used, the sink must be ordered as a five-hole design.

COMPLETION

12. A toilet has a maximum of _____ gpf. Some states mandate a maximum of _____ gpf.

13. The most common distance at which a toilet is installed from a rear wall is _____".

14. The two toilet bowl designs are _____ and _____. A toilet seat must be the same shape as the bowl design.

15. A bathroom sink is known as a(n) _____.

16. The most common faucet hole spread of a lavatory is _____".

17. The two most common laundry sink designs are _____ and _____.

18. A _____ is a personal hygiene fixture adjacent to a toilet.

19. A laundry sink is also called a(n) _____ or _____.

20. The faucet hole spread of a kitchen sink is typically _____" from the hot and cold faucet connections.

Name: _____   Date: _____

Job Sheet 1: Lavatory Ordering Information

- Upon completion of this job sheet, you will be able to demonstrate your understanding on ordering a lavatory.

1. How are lavatories ordered?

2. True or False: A typical residential home utilizes a drop-in style lavatory that is either round or oval.

3. True or False: A drop-in-type lavatory sink requires a typical-size hole cut into the countertop for the particular sink to be installed into the hole.

INSTRUCTOR'S RESPONSE

Name: _____ Date: _____

Job Sheet 2: Bathtubs

- Upon completion of this job sheet, you will be able to demonstrate your understanding about bathtubs.

1. True or False: A standard tub in a home is 6' in length and averages 30" wide.

2. True or False: The depth of water a tub can hold is typically 30'.

3. True or False: Some tubs are sold separately, while others are sold with wall kits as a one-piece tub and shower unit or with numerous whirlpool features.

4. True or False: A tub is typically installed during the trim-out phase of a project.

INSTRUCTOR'S RESPONSE

Name: _____   Date: _____

Job Sheet 3: ADA Requirements for a Toilet

- Upon completion of this job sheet, you will be able to demonstrate your understanding of the Americans with Disabilities Act (ADA) requirements for toilets.

  1. True or False: The tank handle of an ADA-compliant toilet must be located on the side of the tank that has the greatest distance from a sidewall.

  2. True or False: A handle located on the top of a tank typically satisfies ADA handle location regulations.

  3. True or False: The height of a toilet bowl from the floor, which includes the seat, is not regulated by code.

  4. True or False: ADA codes dictate that the minimum height from a floor to a toilet seat is 14-1/2"; the maximum is 17-1/2".

INSTRUCTOR'S RESPONSE

Name: _____   Date: _____

Job Sheet 4: Concrete Floor Penetrations

- Upon completion of this job sheet, you should be able to discuss the local plumbing code in your area regarding concrete floor penetrations.

PROCEDURE

Research the local plumbing codes in your area and, in the space provided below, state the code requirements for minimum front clearance of a toilet.

INSTRUCTOR'S RESPONSE

Faucets and Drain Assemblies

CHAPTER 7

OBJECTIVES
Upon completion of this chapter, you will be able to:

Knowledge-Based
- Understand the differences in basic faucet designs.
- Recognize various faucet styles and finishes.
- Identify the variations in fixture outlets and drain assemblies.

Skill-Based
- Order the correct faucet and drain assembly for a particular fixture.

KEYWORDS

aerator
BW&O
custom home
diverter
escutcheon
finish
knock-out
pop-up
port opening
spec home
T&S faucet
trim

INTRODUCTION

The final faucet selection will frequently be made by the homeowner. During this process, a plumber must consider several things:

- Selecting which manufacturer's product to install is based on cost, quality, and preferred faucet design.
- Most master bathrooms and guest bathrooms have more expensive finishes than other bathrooms in a house.
- All faucet finishes dictate the finishes used for drain assemblies and bathroom accessories to create a color theme.
- Faucet accessories are available to create various themes.
- A faucet installed on a sink through a countertop or through a tub platform is considered a deck-mounted faucet.
- All faucets must be designed to prevent backflow of wastewater into the water distribution system.
- All codes dictate that a faucet must have an air gap or be protected with a vacuum breaker or an approved check valve.

CHAPTER REVIEW QUESTIONS AND EXERCISES

SHORT ANSWER

1. What is the purpose of the drain assembly?

   _____
   _____
   _____

2. What is the purpose of the air gap?

   _____
   _____
   _____

3. What is the main requirement for selecting a kitchen sink faucet?

   _____
   _____
   _____

TRUE/FALSE

4. _____ The kitchen sink drain assembly does not require a basket strainer. Explain your answer.

5. _____ A typical three-hole lavatory sink has a spread of 6" and is often referred to as a center set. Explain your answer.

6. _____ All lavatory sinks have an overflow slot. Explain your answer.

7. _____ All shower drains require a safety pan to be installed. Explain your answer.

8. _____ A laundry sink either will have a drain connection manufactured as part of the sink, or will use a junior basket strainer. Explain your answer.

9. _____ If the hygiene spray is below the flood level rim of the bidet, then a vacuum breaker or other backflow prevention device must be installed. Explain your answer.

Job Sheet 1: Faucet and Drain Questions

- Upon completion of this job sheet, you should be able to identify the common faucet and drain assemblies.

    1. A(n) _____ is a flange installed to conceal pipe penetrations through a wall, floor, or ceiling.

    2. _____ is the color or polish of a faucet, drain assembly, or other fixture trim item.

    3. A tub faucet is mounted either on the _____ or on the _____.

    4. _____ faucets are installed with large capacity tubs, such as a garden tub or whirlpool.

Match the following terms:

a. Escutcheon   _____   5. An opening in a fixture, such as a drain or overflow hole that receives drain assemblies, used to connect the fixture to the drain system.

b. Finish   _____   6. Drain assembly for lavatory sinks and bidets; abbreviated as P.O.

c. Port opening   _____   7. The color or polish of a faucet, drain assembly, or other fixture trim item.

d. Pop-up   _____   8. Flange installed around a pipe to conceal pipe penetrations through a wall, floor, or ceiling.

INSTRUCTOR'S RESPONSE

Name: _____   Date: _____

Job Sheet 2: Faucet Identification

- Upon completion of this job sheet, you should be able to identify the various parts of a faucet.

Identify the various parts of the faucet shown below.

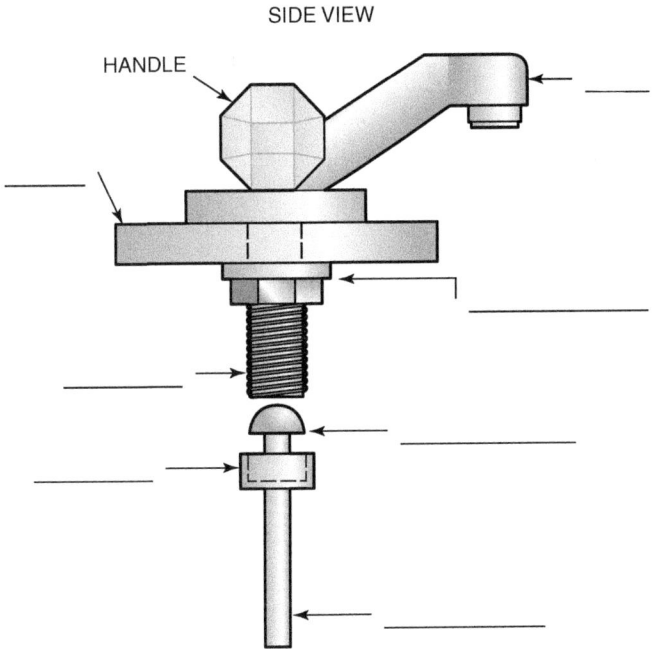

INSTRUCTOR'S RESPONSE

CHAPTER 7 Faucets and Drain Assemblies 73

Name: _____ Date: _____

Job Sheet 3: Bidet Drain Assembly

- Upon completion of this job sheet, you should be able to identify the various parts of a bidet drain assembly.

Identify the various parts of the bidet drain assembly shown below.

SIDE VIEW

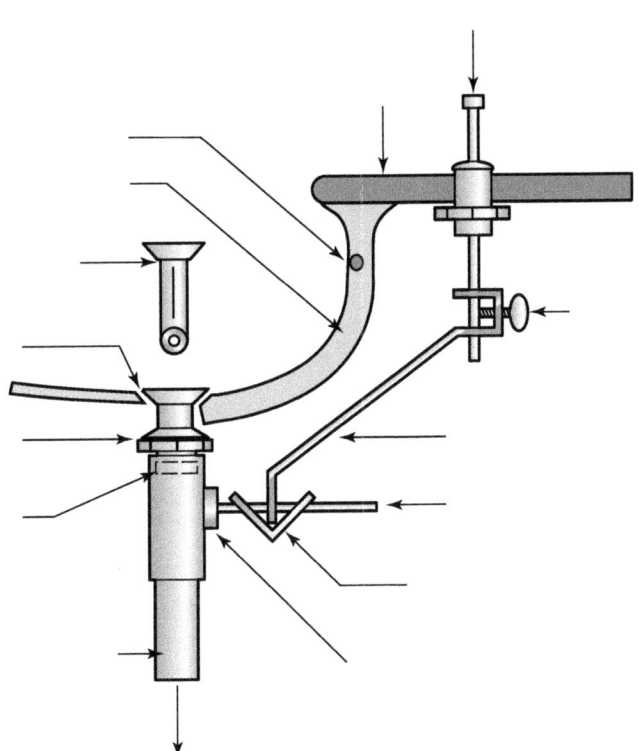

INSTRUCTOR'S RESPONSE

Plumbing Equipment and Appliances

CHAPTER 8

OBJECTIVES

Upon completion of this chapter, you will be able to:

Knowledge-Based
- Explain the differences in water heater designs.
- Understand the basic principles of heating water.
- Explain the variations in equipment connections.

Skill-Based
- Correctly order plumbing equipment.

KEYWORDS

anode rod
burner
collector
element
expansion tank

flue
high limit
instantaneous
pilot
point-of-use water heater

pressure-relief valve
tankless
terminal
thermocouple
thermostat

INTRODUCTION

We can thank plumbing appliances and equipment for making our modern lives much more comfortable. Currently, plumbing equipment and appliances are responsible for washing our clothes and dishes, disposing of waste food products, and heating our water. In some cases, they are even responsible for heating our environment.

GARBAGE DISPOSERS

- Garbage disposers connect to the kitchen sink where the basket strainer would have been installed.
- The horsepower (hp) of the electric motor used by the garbage disposer determines the capabilities of the garbage disposer.
- The most common hp sizes for residential applications range from 1/3 to 3/4.
- Plumbers install a garbage disposer onto a kitchen sink with a specially designed mounting assembly that is unique to each manufacturer.

DISHWASHERS

- Dishwashers are installed by a plumber during the trim-out phase of construction.
- Normally the dishwasher is provided by the homeowner.
- A dishwasher receives its hot water from the same source that services the sink.
- The drain hose from the dishwasher is routed either to the garbage disposer or to a tailpiece specially designed for that purpose.
- The water supply to the dishwasher is typically 3/8" OD tubing routed from the sink to the supply line connection point under the dishwasher.

WASHING MACHINE BOX

- Washing machines are not installed by plumbers; however, plumbers do install the washing machine boxes that provide the water and drain for the washing machine.
- The washing machine box is installed during the rough-in phase of construction.
- It is installed between two vertical wall studs.

ICEMAKER BOX

- Like washing machine boxes, icemaker boxes are installed between two vertical wall studs.
- During the installation process of an icemaker box, the plumber installs a 1/2" pipe to the icemaker box's angle valve.
- The outlet of the angle valve has a 1/4" OD compression connection to allow the compatible tubing of the refrigerator to connect with the icemaker valve.

RESIDENTIAL WATER HEATERS

- Most residential water heaters use either electricity or gas as their energy source.
- Water heaters are equipped with an adjustable thermostat that controls the temperature of the water by starting the heating cycle when the water temperature in the water heater drops below the desired (set) temperature.
- Most manufacturers offer three different versions of their waters: short, standard, and tall.

CHAPTER REVIEW QUESTIONS AND EXERCISES

COMPLETION

1. _____FLUE_____ is the entire pipe system exhausting fumes from a gas water heater.

2. A(n) _____THERMOSTAT_____ is a regulating device to control the water temperature of a water heater.

3. A device that heats water in a solar water heating system, also called a panel, is a(n) _____COLLECTOR_____.

4. A(n) __ANODE ROD__ is a device installed in a water heater to protect the inside of a storage tank from corrosion.

5. A(n) __ELEMENT__ is an electrical heating device used to internally heat water in an electric water heater.

6. __TANKLESS__ is a type of water heater, also called instantaneous, that does not store water in a storage tank.

7. A(n) ~~T&P~~ __HIGH LIMIT DEVICE__ is a safety device on all water heaters to protect from overheating water.

8. The flame of a gas water heater used to ignite gas entering a burner assembly is called the __PILOT__.

9. The __BURNER__ is the main flame assembly that externally heats water in a gas water heater.

10. __TERMINAL__ describes where an electrical wire connects to a device; also referred to as a post.

11. A(n) __THERMOCOUPLE__ is a heat-sensing device to ensure that a gas water heater pilot flame is ignited.

TRUE/FALSE

12. __F__ Washing machine installations typically have a specially designed wall box that houses the hot and cold water supply connections and a 2-1/2" drain.

13. __T__ Most residential electric water heaters are 240 volt and non-simultaneous.

14. __T__ One British thermal unit (BTU) is the amount of heat required to raise 1 pound of water by 1 degree F.

CHAPTER 8   Plumbing Equipment and Appliances   79

Name: _____   Date: _____

Job Sheet 1: Garbage Disposal Identification

- Upon completion of this job sheet, you should be able to identify components used in the installation of a garbage disposal.

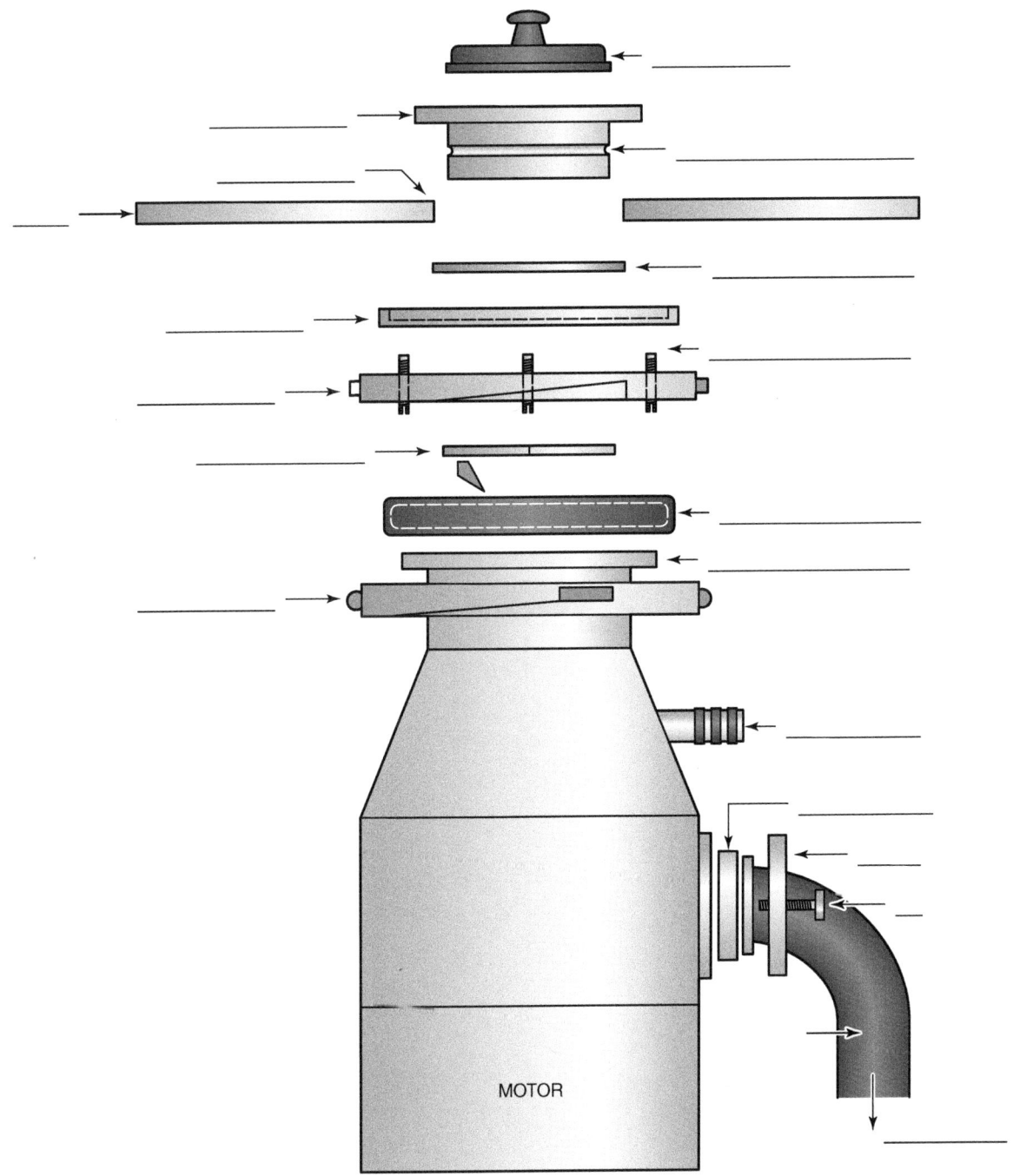

INSTRUCTOR'S RESPONSE

Name: _____   Date: _____

Job Sheet 2: Tankless Water Heater

- Upon completion of this job sheet, you should be able to describe how a tankless water heater works.

Complete the following diagram.

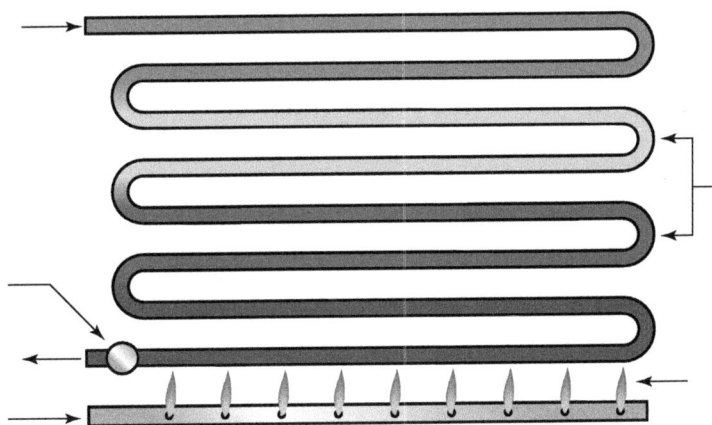

INSTRUCTOR'S RESPONSE

Name: _____   Date: _____

Job Sheet 3: Determining if Food Waste Can Be Discharged into a Septic System

- Upon completion of this job sheet, you should be able to explain if the local plumbing codes allow for food waste to be discharged into a septic tank system in your area.

Research the local plumbing codes in your area to determine if it is permissible to discharge food waste into a septic system. Provide the results of that research below.

INSTRUCTOR'S RESPONSE

Name: _____ Date: _____

Job Sheet 4: Determining the Dishwasher Drain Hose Requirements

- Upon completion of this job sheet, you should be able to explain if the local plumbing code has a requirement for dishwasher drain hose routing (through an air gap device).

Research the local plumbing codes in your area to determine the requirements for a dishwasher drain hose. Does the code dictate the use of an air gap device? Provide the results of that research below.

INSTRUCTOR'S RESPONSE

Blueprint Reading and Drafting

CHAPTER 9

OBJECTIVES

Upon completion of this chapter, you will be able to:

Knowledge-Based
- Understand basic plumbing symbols and abbreviations.
- Understand the different illustrated views of a piping system.

Skill-Based
- Interpret basic residential architectural blueprints.
- Create simple sketches of piping systems.

KEYWORDS

drafting triangle
drawings
isometric view
joist

load-bearing wall
plan view
riser diagram
section view

side view
sketch
stud
tee

INTRODUCTION

To become an effective plumber, you must have a good understanding of blueprint reading and basic drawing techniques. A plumber is typically not responsible for creating plumbing drawing, but will be required to follow plumbing drawings for bidding and installation.

DRAFTING TOOLS

- Like all other building construction trades, drafting employs specialty tools to produce quality and accurate drawings.
- Plumbers are not required nor are they expected to purchase expensive drafting tools to produce sketches.
- As you get more practice creating freehand sketches, your drawings will become more effective.

SCALE RULER

- Scale rulers are used to determine dimensions on a blueprint.
- The two most common scales used for plumbing blueprints are 1/8" and 1/4".
- 6" flat scales are perfect for carrying on to job site because they fit perfectly in a shirt pocket.
- A tape measure can be used as a scale ruler on a job site.

DRAFTING TRIANGLES

- Drafting triangles are available in various sizes. The two most commonly used drafting triangles are the 45° and the 30° to 60° triangles.
- 30° to 60° triangles are commonly used for creating isometrics.
- 45° triangles are commonly used for creating plan and side views.

SYMBOL TEMPLATES

- Symbol templates can be used to quickly create standard symbols that are uniform in appearance and require little skill.
- Templates range from circular templates (used to create circles) to identification templates (used to create revision symbols, identification symbols, etc.) to specialty templates.

DRAFTING PAPER

- Drafting paper comes in a variety of different sizes and types.
- Grid paper can be used to aid in creating sketches and is used by those beginning to develop basic drafting skills.
- Vellum is a type of drafting paper available in a range of sizes.
- Vellum is specially designed to reproduce blueprints.

ISOMETRIC DRAFTING

- Isometric views of piping systems are used to clarify piping configurations that are difficult to visualize in any other view.
- Isometric views are typically not drawn to scale; however, dimensions can be added to show fabrication intent or distances.

RISER DIAGRAMS

- An isometric drawing that illustrates vertical piping that extends through multiple floors is known as a riser diagram.

Chapter Review Questions and Exercises

TRUE/FALSE

1. __T__ Common plumbing symbols indicate the intent of a design.
2. __T__ Abbreviations are used on a blueprint to eliminate clutter and are usually listed on a legend with their meanings.
3. __F__ Most residential blueprints do illustrate the pipe routes for plumbing systems.
4. __T__ An architect typically only illustrates the fixture locations on a residential blueprint.
5. __T__ Drafting skills allow a plumber to effectively communicate a design to co-workers and other tradespersons.
6. __F__ A plan view is not also known as a birds-eye view.
7. __T__ A side view illustrates an area of construction from the side.
8. __T__ An isometric view is a three-dimensional view of a piping system.

SHORT ANSWER

9. What is a riser diagram?

   _____
   _____
   _____
   _____

10. What are the two common triangles used in drafting?

    _____
    _____
    _____
    _____

Name: _____ Date: _____

Job Sheet 1: Residential Plumbing Symbols

- Upon completion of this job sheet, you should be able to draw residential plumbing symbols.

In the space provided below draw the following symbols.

1. Perpendicular tee configuration

2. Tee

3. 45° offset

4. 90° offset

5. p-trap

INSTRUCTOR'S RESPONSE

Name: _____ Date: _____

Job Sheet 2: Using a Tape Measure for a Scale Ruler

- Upon completion of this job sheet, you should be able to use a tape measure as a scale ruler.

In the space provided below draw the following at their indicated scale.

1. 4'-6" (Scale 1/4")

2. 8'-0" (Scale 1/8")

3. 6'-3" (Scale 1/4")

4. 12'-0" (Scale 1/4")

5. 7'-6" (Scale 1/8")

INSTRUCTOR'S RESPONSE

Name: _____   Date: _____

Job Sheet 3: Creating a Riser Diagram

- Upon completion of this job sheet, you should be able to create a riser diagram.

Using Figure 9-44 in *RCA Plumbing*, 2nd edition, redraw the riser diagram.

INSTRUCTOR'S RESPONSE

Material Organization and Layout

CHAPTER 10

OBJECTIVES

Upon completion of this chapter, you will be able to:

Knowledge-Based
- Respect material organization methods to increase productivity.
- Know that the layout and installation of a plumbing system differs for each project.

Skill-Based
- Demonstrate safe material-handling techniques.
- Use manufacturer installation data properly.
- Demonstrate proper system layout.

KEYWORDS

carpenter
joist
load-bearing

procure
slab

stud
trench

INTRODUCTION

A construction site operates only when all trades provide effective, proper communication. The two forms of communication are written and oral. Written communication is the most reliable, but if information is required immediately, oral can be more productive. A plumber reviewing a blueprint is communicating a design from the architect.

COMMUNICATION

- Communication is an important aspect of any job, especially in a residential construction job site.
- This includes both written and verbal communications.

WRITTEN COMMUNICATION

- Written communication is the most effective form of communication on a job site today.
- Each employer must have a Hazardous Communication Program, in which information regarding dangers and safety procedures to employees are disseminated.
- OSHA requires that a Material Safety Data Sheet (MSDS) be made available upon request for every product that contains hazardous or dangerous materials.
- In addition to indicating the danger to exposure to certain materials, the MSDS indicates the necessary medical treatment necessary for such exposure.

ORAL COMMUNICATION

- Oral communications are one of the leading forms of errors on a construction site.
- For single-family dwelling construction sites the plumber and an apprentice will install the entire plumbing system.
- Typically for a single-family residential construction site the architectural blueprints will not show the piping system.
- Advances in technology have made it possible for a plumber to have immediate contact from a job site to a supervisor.
- Many mobile phones have the ability to be used as two-way radios.
- Smaller construction sites may use two-way radios for verbal communications within a smaller radius.

MATERIAL ORGANIZATION

- Material organization is critical if a task is to be completed in a productive manner.
- Without material organization skills a plumber may not order all the necessary materials for a particular job.
- Before materials can be organized the plumber must have a working knowledge of the task the materials are to be used for.
- Plumbing companies usually provide their plumbers with extra materials before the plumber goes to the job site.
- The potential savings in labor offset the additional materials cost, making this justifiable.
- One way in which to reduce cost on a construction site is to make a list of the needed materials before starting a particular job.
- Methods of material organization that can be used on a residential construction site include:
  – Creating a materials list
  – Palletizing material
  – Bagging and tagging
  – Warehousing on or off site
  – Procuring material per job site or installation

CHAPTER REVIEW QUESTIONS AND EXERCISES

COMPLETION

1. A(n) _____ is a vertical board used to erect a wall.

2. A(n) _____ is a portion of the structure that bears the weight of the structure, such as a load-bearing wall.

3. A(n) _____ is an individual installing the wood framing or other woodwork.

4. What are the two forms of communication?

   _____

5. _____ is the process of receiving material through ordering or gathering material.

6. In many southern regions of the country, _____ or _____ designs are common; northern regions employ more _____ foundation systems.

7. A(n) _____ is a concrete floor used to define a building design that does not have a crawlspace or basement.

8. A(n) _____ is the horizontal board used for structural integrity.

TRUE/FALSE

9. _____ Material organization skills are essential in completing a task in a productive manner. Explain your answer.

10. _____ Rubber straps can break and cause injury, especially in hot climates. Explain your answer.

11. _____ Bagging and tagging is a method used to organize material for specific tasks. Explain your answer.

12. _____ Layout of a piping system must include fixture types and various plumbing codes. Explain your answer.

13. _____ The size of a hole and the location a hole can be drilled are the same regardless of the load-carrying responsibility of the wall stud. Explain your answer.

Name: _____   Date: _____

Job Sheet 1: Material Organization

- Upon completion of this job sheet, you should be able to identify several methods for organizing materials.

List the methods for organizing job-site materials.

INSTRUCTOR'S RESPONSE

Name: _____   Date: _____

Job Sheet 2: Concrete Floor Penetrations

- Upon completion of this job sheet, you should be able to discuss the local plumbing code in your area regarding concrete floor penetrations.

PROCEDURE

Research the local plumbing codes in your area and in the space provided below state the code requirements for piping that penetrates concrete floors.

INSTRUCTOR'S RESPONSE

Water Service Installation

CHAPTER 11

OBJECTIVES

Upon completion of this chapter, you will be able to:

Knowledge-Based
- Understand correct installation techniques for installing a water service.
- Understand the basics of municipal and private water systems.
- Understand basic codes pertaining to burial depths and locations.
- Understand and respect water quality issues and regulations.

KEYWORDS

- aquifers
- backfill
- brackish
- branch
- compacting
- filter
- purification
- trench
- water distribution system
- water service

INTRODUCTION

Potable water is a primary concern of most plumbing jobs. Without an adequate potable water supply, it becomes impossible to maintain many of the modern conveniences and services that are necessary to sustain the quality of life that our society has grown accustomed to. Potable water is used not only for drinking water but for any application that might come in contact with humans.

WATER SOURCE

- Water that is safe for human consumption is often referred to as drinking or domestic water.
- Homes that utilize a well pump system get their water from the natural water tables below the surface.
- Water that is below and above the ground is replenished by rain water and natural underground springs.

PUBLIC WATER SYSTEM

- A public water system is defined by the Environmental Protection Agency (EPA) as a system that is in service for a minimum of 60 days per year, provides water to a minimum of 15 connections or 25 individuals.
- Public water systems are regulated by the federal Safe Drinking Water Act (SDWA).
- Water towers are used in a domestic water system to provide pressure to the piping system.
- One vertical foot of water exerts 0.433 psi of pressure.
- 100 vertical feet of water would exert 43.3 psi of pressure.

PRIVATE WATER SYSTEMS

- Water systems that do not service the minimum EPA guidelines for a public water system are known as private water systems.
- Ground water is the most common source of water in a private water system.
- There are two basic classifications of wells: shallow wells and deep wells.
- Shallow wells are less than 50 feet in depth, while deep wells can be in excess of 1000 feet in depth.

EPA STANDARDS

- Typically it is the local health administration that regulates and inspects the purity of water regardless of its source.
- New California regulations prohibit the use of products containing lead in any system with drinking water.

WATER QUALITY

- EPA drinking water standards are categorized in two areas of regulation: Natural Primary Drinking Water Regulations and Natural Secondary Drinking Water Regulations.
- The Natural Primary Drinking Water Regulations is the primary standard that is enforceable by law.

WATER FILTRATION

- Color, odor, and taste can be improved with the use of water filtration.
- Filtration is used to remove sediments as well as contaminants.
- Chlorine, fluoride, and other necessary chemicals are added during the purification process.
- Nitrate and coliform bacteria are two common threats for private water systems.

Chapter Review Questions and Exercises

COMPLETION

1. _____ is the piping from a water meter or well to a building and connects to the water distribution system.

2. A(n) _____ is a geologic formation containing water.

3. Lowland water close to the ocean containing high salt content is said to be _____.

4. _____ is loose soil placed into an excavated area; also called fill.

5. A(n) _____ is a pipe installed laterally from a main pipe.

6. _____ is the process of compressing loose soil placed back in a trench; also known as tamping.

7. _____ is the entire system for distributing water; this refers to the piping inside a house.

8. A(n) _____ is an accessory to remove particulate from water, but does not purify the water.

9. _____ is the process to cleanse the water to ensure it is potable.

10. A(n) _____ is an excavated pocket of soil to install piping; also known as a ditch.

TRUE/FALSE

11. _____ Segment identification of certain drains and vents is required to properly size a drainage waste and vent (DWV) system.

12. _____ A drainage fixture unit (dfu) is the basis for sizing a DWV system.

13. _____ A code book is not required to locate the allowable dfu load on a particular pipe size.

14. _____ Slope is required on a horizontal drain and varies based on the pipe size.

15. _____ California enacted a new law redefining "lead free" on January 1, 2008.

Name: _____   Date: _____

Job Sheet 1: Water Meter Installation for Cold Weather Climates

- Upon completion of this job sheet, you should be able to identify the components of a cold weather water meter installation.

Identify the various components of this installation.

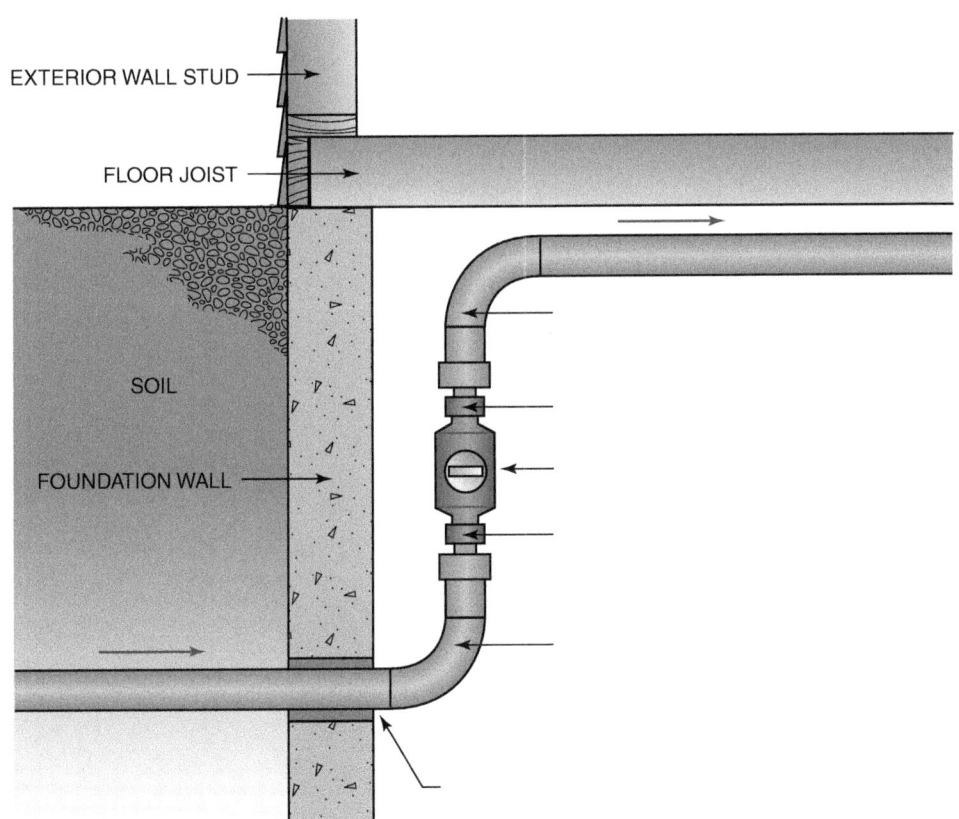

INSTRUCTOR'S RESPONSE

Name: _____   Date: _____

Job Sheet 2: Determining the Code Requirements for Working on a Public Water Supply

- Upon completion of this job sheet, you should be able to explain the requirements for working on a public water supply.

PROCEDURE

In the space provided below, answer the following questions: What are the requirements for working on a public water supply as outlined by the plumbing codes in your area? If a special license is required, what are the requirements for that license?

INSTRUCTOR'S RESPONSE

Name: _____ Date: _____

Job Sheet 3: Water Service Installation Code Requirements

- Upon completion of this job sheet, you should be able to explain the minimum allowable water service as dictated by the plumbing codes in your area.

PROCEDURE

In the space provided below, answer the following question: What is the minimum allowable water service in your area?

INSTRUCTOR'S RESPONSE

Water Distribution Installation

CHAPTER 12

OBJECTIVES

Upon completion of this chapter, you will be able to:

Knowledge-Based
- Know correct techniques for installing a water distribution system.
- Respect that a plumber installs water piping per code to protect water quality.
- Know drilling and notching codes to ensure the structural safety of a building.
- Understand the rough-in aspects of a water supply to specific fixtures.

KEYWORDS

- brazing
- flux
- International Plumbing Code (IPC)
- joint
- joist
- partition
- silver solder
- solder
- soldering
- stud
- torch
- Uniform Plumbing Code (UPC)
- water supply fixture unit (wsfu)

INTRODUCTION

The hot and cold water piping in a facility is known as the water distribution system. Water piping is typically installed after the completion of the drainage and vent piping because the water piping system consumes less space in walls and ceilings and has less strict installation code regulations. The installation of water distribution systems is different for commercial and residential in the type of equipment installed as well as the size of pipe used. In addition, some codes that regulate the installation of plumbing systems differ. The sizing of an entire system is established based on the quantity and type of fixtures being served. The three main segments of a water distribution piping system are the main, branch main, and individual supply.

LAYOUT AND SIZING

- The layout process starts by first determining the location of each fixture, their requirements, and the location of the water service entering the residence as well as the route of the main piping system.

PIPE SIZING

- The quantity and type of fixtures being served in a residence is used to establish the systems base.
- Plumbing codes provide maximum number of fixtures that can be served on a specific pipe size.
- The piping system is based on the maximum GPM or GPF of a particular fixture and then calculated as per all fixtures.
- The International Plumbing Code (IPC) and the Uniform Plumbing Code (UPC) differ slightly in their pipe sizing allowances.

SIZING THEORY

- Determining the available water of a certain pipe size is based on volume and pressure.
- The rate at which water flows through the piping is known as velocity.
- A water supply fixture unit (wsfu) is the designated term used to describe the end result of a flow calculation.
- The amount of water supplied to a fixture is determined by the pipe size.
- The speed at which water is delivered to a pipe is determined by the pressure.

JOB SITE SIZING

- An isolation valve known as a stop is used on toilets and sinks.
  - The most common stop connects 1/2" (5/8" OD) pipe to 3/8" OD tubing.
  - The size of the stop to be installed serving a fixture often determines the size of the rough-in water pipe more than the code.
- The specific fixture dictates the termination of an individual supply pipe.
  - Most codes dictate that the individual supply piping must be routed to a maximum of 30" from the fixture it serves to ensure adequate water flow at the fixture.

CHAPTER REVIEW QUESTIONS AND EXERCISES

TRUE/FALSE

1. _____ The layout of a water distribution system varies based on the actual job site and the fixtures being served.

2. _____ The sizing of an entire system is established based on the quantity and type of fixtures being served.

3. _____ Installing a combination tub and shower style faucet will change the sizing requirements of the individual fixture supply.

4. _____ Plumbers are not concerned with drilling and notching codes. Typically that is the concern of the framing crew.

5. _____ Soldering copper must be performed with approved lead-free products.

6. _____ Follow manufacturer instructions when performing a solvent weld connection.

7. _____ Currently plumbing codes do not dictate the maximum hanger and support spacing.

COMPLETION

8. A(n) _____ is the designated term to describe the end result of a flow calculation.

9. One gallon per minute is _____ liters per second.

10. The rate the water flows through the piping is known as _____ and is measured in feet per second.

11. The wall rough-in for a water distribution system is determined by the _____.

12. The standard water rough-in dimension for a tank type toilet is _____" to the left of the toilet center.

13. The maximum height above the finished floor for most lavatory sink water piping regardless of the type is _____".

14. Hanger and support selection is based on _____.

15. A(n) _____ is a vertical board used to create a wall and is categorized as load-bearing and non-load-bearing.

16. _____ is the welding process used without flux to weld copper tube.

17. _____ is the process of welding copper tube by using flux and solder.

18. A horizontal board that is part of a complete framing is a(n) _____.

19. A(n) _____ is a fitting connection for a pipe.

20. A non-load-bearing wall designed for separating rooms and not to support a structural load is a(n) _____.

21. A(n) _____ is a tool that is ignited and creates a flame to solder or braze copper tube.

22. _____ is a metal filler used to solder copper tube; it cannot contain more than 0.2% lead and is in roll form.

23. _____ is a chemical paste used to solder copper tube.

Name: _____   Date: _____

Job Sheet 1: Brazing

- Upon completion of this job sheet, you will be able to discuss and perform a brazing exercise.

1. Using the *RCA Plumbing* 2nd edition textbook, outline the procedure for brazing copper tubing.

Now you try it.

Step A. Using a tape measure and a tubing cutter, measure and cut two pieces of 1/2" copper tubing 12" long.

Step B. Ream the end of the copper tubing.

Step C. Using a 1/2" coupling assemble the two pieces of copper tubing and coupling together.

Step D. Check with your instructor before igniting the torch; braze the assembly together.

Step E. Place the flame over the pipe to heat that area first and then heat the fitting socket all the while moving the flame back and forth to each area and around the entire fitting.

Step F. Place the silver-solder rod on the fitting socket where it intersects with the pipe.

Step G. Clean work area and tools.

Step H. Return all tools to their proper location.

Step I. Have your instructor inspect the connections and work area.

INSTRUCTOR'S RESPONSE

Name: _____   Date: _____

Job Sheet 2: Soldering

- Upon completion of this job sheet, you will be able to discuss and perform a soldering exercise.

    1. Using the *RCA Plumbing* 2nd edition textbook, outline the procedure for soldering copper tubing.

Now you try it.

Step A. Using a tape measure and a tubing cutter, measure and cut two pieces of 1/2" copper tubing 12" long.

Step B. Ream and wipe off the end of the copper tubing.

Step C. Clean the inside of the coupling.

Step D. Insert a wire brush inside the fitting and rotate it clockwise several times until the fitting is visibly clean.

Step E. Apply flux to the clean, dry, oil-free pipe ends.

Step F. Apply a thin layer of flux into the clean, dry, oil-free fitting socket.

Step G. Using a 1/2" coupling assemble the two pieces of copper tubing and coupling together.

Step H. Check with your instructor before igniting the torch; braze the assembly together.

Step I. Place the blue portion of the flame on the rear of the fitting socket.

Step J. Move the torch to evenly distribute the heat.

Step K. Remove the heat from the pipe, and wipe the excess flux with a clean, dry rag, as necessary.

Step L. Apply the heat again, and place the solder onto the edge of the fitting where it connects with the copper tube.

Step M. Clean work area and tools.

Step N. Return all tools to their proper location.

Step O. Have your instructor inspect the connections, and work area.

INSTRUCTOR'S RESPONSE

Name: _____ Date: _____

Job Sheet 3: Flaring Copper Tubing

- Upon completion of this job sheet, you will be able to discuss and flare copper tubing.

    1. Using the *RCA Plumbing* 2nd edition textbook outline the procedure for flaring copper tubing.

Now you try it.

Step A.	Using a tape measure and a tubing cutter, measure and cut two pieces of 1/2" copper tubing 12" long.
Step B.	Ream and wipe off the end of the copper tubing.
Step C.	Open (spread) the clamp portion of the flaring tool, and insert tubing into the correct hole of the tool.
Step D.	Install the flaring portion of the tool over the clamp portion.
Step E.	Turn the flaring tool handle clockwise, so the flaring post to the end of the tube is flared.
Step F.	Remove the tool and slide the flare nut over the newly created flare to make sure the flare is the right diameter.
Step G.	Clean work area and tools.
Step H.	Return all tools to their proper location.
Step I.	Have your instructor inspect the connections and work area.

INSTRUCTOR'S RESPONSE

Drainage Waste and Vent Segments and Sizing

CHAPTER 13

OBJECTIVES
Upon completion of this chapter, you will be able to:

Knowledge-Based
- Identify and describe segments of a drainage waste and vent (DWV) system.
- Recall the basic abbreviations of a DWV system.
- Understand basic conventional septic system operation.

Skill-Based
- Size the various segments of a DWV system.

KEYWORDS

- air admittance valve (AAV)
- branch
- branch interval
- branch vent
- building drain
- building sewer
- circuit vent
- cleanout
- common vent
- D-box
- developed length
- distribution box
- drain
- drain field
- drainage fixture unit (dfu)
- drainage waste and vent (DWV)
- effluent
- fixture branch
- fixture drain
- horizontal branch
- hydraulic gradient
- individual vent
- interval
- loop vent
- open air
- p-trap
- perc test
- percolation
- relief vent
- rough-in
- septic tank
- soil stack
- stack
- stack vent
- trap
- trap adapter
- trap distance
- vent stack
- waste stack
- wastewater
- weir

INTRODUCTION

Before a drainage waste and vent (DWV) system can be installed, it must be properly designed and sized. The size requirements for a DWV are regulated by the state in which the system is being installed. Designing and installing a DWV system can be a complex and extensive process. Pipe routes are a result of determining fixture locations, fixture requirements, relative codes, construction obstacles, coordination with other trades, and company installation standards. Every plumbing fixture connected to a drainage system must be protected by a fitting or device known as a trap. Always consult your local building codes.

Major Segments of a DWV System

The major segments of a DWV system

- BD: Building Drain
- BS: Building Sewer
- CO: Cleanout
- SV: Stack Vent
- VS: Vent Stack
- VTR: Vent Through Roof
- WS: Waste Stack

BUILDING DRAIN

- The building drain is the lowest horizontal portion of a drainage system and receives discharge from waste stacks and horizontal branches.

WASTE STACK

- The waste stack is the main vertical pipe that begins with its connection to the building drain and terminates with its connection to the stack vent.

STACK VENT

- The vent for the waste stack is known as the stack vent and begins at the highest branch connection to the waste stack.

VENT STACK

- It is sized based on numerous factors including the total discharge load of a system and the length it travels.

CLEANOUT

- All codes dictate that the base of every stack and the transition from a building drain and building sewer must have a cleanout installed.

Minor Segments of a DWV System

The minor segments of a DWV system

- BV: Branch Vent
- CV: Circuit Vent
- FB: Fixture Branch

- FD: Fixture Drain
- HB: Horizontal Branch
- IV: Individual Vent
- LV: Loop Vent
- RV: Relief Vent

Chapter Review Questions and Exercises

COMPLETION

1. A(n) _____ is the lowest horizontal main drain of a DWV system; it conveys wastewater to a building sewer.

2. A one-way valve that allows air to enter a DWV system is used in place of a vent that would normally terminate with another vent or through a roof, and is abbreviated as AAV is a(n) _____.

3. _____ is the section of piping of a DWV system that connects to main portions of a system.

4. A(n) _____ is a nonrestrictive fitting installed at each fixture that does not have an integral trap and utilizes a water seal to eliminate sewer gases from entering occupied areas; often referred to as a trap.

5. _____ is the vertical distance along a stack equal to one story height, but no less than 8'-0" and the area where horizontal branches connect to a stack.

6. A(n) _____ is a fitting used to connect tubular piping to other pipe connections.

7. A fabricated box or structure that distributes effluent to a drain field or other designated location is a(n) _____.

8. A(n) _____ is the complete system of draining soil, waste, and wastewater to a point of disposal and circulating air within the system.

9. A(n) _____ is the area of installation for perforated piping to drain wastewater (effluent); also called a leach field.

10. The access point to remove obstructions from a DWV system is a _____.

11. A(n) _____ is the vent serving one fixture trap; it terminates to open air or connects with another vent.

12. A(n) _____ is the vertical vent pipe that receives other vents and terminates or uses a stack vent to open air; can be installed with horizontal offsets.

13. A pipe used to circulate air between a drainage and vent system is a(n) _____. It has several specific areas of installation and is sized based on its use.

14. A(n) _____ is the dimension a trap weir is located from its protective vent.

15. The phase of construction before the finish or trim phase when all piping is installed in floors, walls, and ceilings is the _____ phase.

16. _____ is the water that does not contain sewage and the term often used by plumbers instead of the word "effluent."

17. A(n) _____ is the vertical pipe conveying wastewater only; it is the same pipe as a soil stack, and can be installed with horizontal offsets.

18. _____ is the finish phase of construction when all fixtures are installed; also known simply as trim.

19. _____ is wastewater that does not contain fecal matter.

Name: _____ Date: _____

Job Sheet 1: Identify the Major Segments of a DWV System

- Upon completion of this job sheet, you should be able to identify the major segments of a DWV system. In the space provided below, list and describe the major segments of a DWV system.

INSTRUCTOR'S RESPONSE

Name: _____ Date: _____

Job Sheet 2: Identify the Minor Segments of a DWV System

- Upon completion of this job sheet, you should be able to identify the minor segments of a DWV system.

In the space provided below, list and describe the minor segments of a DWV system.

INSTRUCTOR'S RESPONSE

Drainage Waste and Vent Installation

CHAPTER 14

OBJECTIVES

Upon completion of this chapter, you will be able to:

Knowledge-Based
- Know basic layout considerations based on common fixture types and codes.
- Be able to recognize pipe route installations based on structural obstacles.
- Apply fitting knowledge learned in other chapters to correct installation practices.
- Know the correct testing methods for passing a plumbing inspection.
- Recognize that company preference can dictate installation practices.

KEYWORDS

fixture group
half bathroom
rough

test ball
test cap
test plug

trim-out
waste arm

INTRODUCTION

A plumber must consider all aspects of a plumbing system route before beginning the layout process. Every job site has unique layout challenges and a plumber begins with identifying and laying out the fixture locations.

The installation of a DWV system is based on codes, fixture types, fixture locations, floor joist direction, and wall stud layout. An air admittance valve (AAV) is allowable by many codes to eliminate installing some vents to terminate to open air. The height at which a vent pipe terminates above a roof varies based on local code. The distance a trap can be installed from its vent is known as trap distance.

SCOPE OF WORK
- The scope of work is the total knowledge of the entire project.
- The type of materials, fixtures, and design must be known.

GUEST BATHROOM LAYOUT
- A bathroom having only a toilet and a sink is known as a half bathroom.
- A full bath has a toilet, lavatory, and bathtub and/or shower.

KITCHEN SINK LAYOUT

- An AAV is commonly installed below the kitchen sink to minimize labor costs.

MASTER BATHROOM

- A plumber must approach a piping route to a bathroom group based on the piping requirement serving a toilet.

HALL BATHROOM

- Most codes dictate that a joist cannot be drilled in the top and bottom 2".
- Many codes do not allow large diameter holes in the middle one-third of the joist span.
- Most codes dictate that a 2" pipe must slope 1/4" per foot.

LAUNDRY ROOM

- Most codes state that the minimum distance is 18" and the maximum is 42" of the standpipe from a washing machine box to the inlet of a p-trap.
- Some codes do not allow a p-trap serving a washing machine to be installed below floor level.
- Many codes require the piping to be increased to 3" even though the p-trap can remain 2".

BUILDING DRAIN

- A building drain is the lowest horizontal portion of a DWV system.

VENTING SYSTEM

- Most codes dictate that at least one vent in a system must be a minimum of 3" in size throughout the entire length of that vent.

FIXTURE ROUGH-IN

- The drainage and vent systems are typically installed before the water piping due to the physical size of DWV piping being larger and hole size requirements.
- A 1-1/4" pipe is the smallest size allowable in a DWV system, but most contractors do not install piping smaller than 1-1/2".

TOILETS

- There are two basic variables in the rough-in of the drainage piping relating to the distance from the back wall.
- The two most common rough-in dimensions are 10" and 12".

Chapter Review Questions and Exercises

COMPLETION

1. A(n) __WASTE ARM__ is a horizontal fixture drain that begins with the vent connection and terminates at the fixture connection.

2. A bathroom consisting of a tub or shower, toilet, and a lavatory is called a __BATHROOM (FIXTURE) GROUP__.

3. __TEST CAPS, TEST BALL, FERNCO CAP__ are used to seal a pipe for testing a DWV system.

4. __FINISH PHASE / TRIM OUT__ is the finish phase of construction when fixtures are installed.

5. __TEST BALL__ is a testing accessory that is injected with air after being inserted into a drainage pipe to allow the system to be filled with water.

6. A testing accessory inserted into a drainage pipe and injected with air, allowing the system to be filled with water, is a(n) __TEST BALL / PLUG__.

7. A(n) __HALF BATH / POWDER ROOM__ is a bathroom consisting of only a toilet and a lavatory.

8. The rough-in dimension from the back wall to the center of the toilet drain is called the __ROUGH__.

Name: _____   Date: _____

Job Sheet 1: Common DWV Layout Considerations

- Upon completion of this job sheet, you should be able to identify the considerations for laying out a DWV system.
- This job sheet is a checklist of the basic considerations for laying out a DWV system.

Consideration	Notes	Comments	Completed By
Sewer entry location	Which side of the house?		
Number of bathrooms	Total fixtures and location of each bathroom in house		
Bathroom fixture layout	Fixture relation to each other and types		
Kitchen sink layout	Location in house and if garbage disposer and DW are being installed		
Washing machine layout	Location in house		
Wall relation to bathroom groups and other fixtures	Wall types, sizes, and location of wall studs		
Floor joist direction and relation to bathroom groups	In relation to bathtub and toilet fixture drain requirements		
Fixture types	Specific fixture rough-in requirements		
Venting code allowances	If AAV and wet venting are allowed		
Vent terminations	Penetration locations through roof		

INSTRUCTOR'S RESPONSE

Name: _____   Date: _____

Job Sheet 2: Wet Venting Code Requirements

- Upon completion of this job sheet, you should be able to discuss the local plumbing code requirement regarding wet venting.

Knowing where the vents terminate through the roof and what fixtures will use AAV is essential for establishing the vent routes. If your local code allows wet venting, actual pipe routes are established for particular groups of fixtures. What are the requirements in your area according to your local plumbing codes regarding wet venting?

INSTRUCTOR'S RESPONSE

Fixture and Equipment Installation

CHAPTER 15

OBJECTIVES

Upon completion of this chapter, you will be able to:

Knowledge-Based
- Know basic fixture and equipment installations.
- Know the tools and materials required before beginning an installation.
- Understand the sequence of installing materials to complete a task in a productive manner.
- Recognize that company preference can dictate installation practices.

KEYWORDS

closet bolts
closet flange
compression
continuous waste
escutcheon
slip joint
trap adapter
tubular
wax seal

INTRODUCTION

Most fixture installations occur during the trim-out phase of construction. The installation of plumbing fixtures varies. For example, a toilet is installed onto a closet flange with a wax seal and a set of closet bolts.

ESCUTCHEONS AND STOPS

- Every ceiling, wall, or floor stub-out must have an escutcheon installed to conceal the pipe penetration.

TOILETS

- The water supply is located on the left side of all toilets.
- The water is connected to a toilet using a tank supply.
- The toilet is installed onto the closet (toilet) flange and sealed with a wax ring.
- Some wax seals have a plastic accessory called a horn molded into the wax.

LAVATORIES

- Most residential lavatory sinks are pre-molded cultured marble that are a single entity with the countertop.
- The stub-out piping serving a lavatory is either 1-1/4" or 1-1/2".

Kitchen Sinks

- The two common types of kitchen sinks used in residential construction are stainless steel and cast iron.
- A plumber applies caulking to the edge of the cutout area of the countertop and places the cast iron sink into the hole.

Laundry Sinks

- The two most common types of laundry sink designs are the following:
  - Wall mounted
  - Four legs secured to the floor

Electric Water Heaters

- An electric water heater only requires a plumber to connect the hot and cold water piping from the rough-in stub-outs to the designated inlet and outlet connection.

Gas Water Heaters

- Codes vary pertaining to the gas supply connections and venting regulations.

Chapter Review Questions and Exercises

COMPLETION

1. _Closet Bolts_ are non-corrosive bolts inserted into a closet flange to anchor the toilet to the drainage system.

2. A(n) _Trap Adapter_ is a type of drainage connection between the rough-in stub-out and tubular p-trap.

3. Another name for a trap adapter is a(n) _Desanco_.

4. A(n) _Closet Flange_, also called a toilet flange, is installed to connect a toilet to the drainage system.

5. A type of connection used for water distribution tubing is called a(n) _Compression Connection_

6. A(n) _Continuous Waste_ is a drain assembly used to connect a double-bowl kitchen sink to a single p-trap.

7. A slip-joint style fitting installed to connect a tubular p-trap to the drainage system is a(n) _Trap Adapter_.

8. A(n) _Escutton_ is a floor or wall plate installed around a stub-out pipe to conceal the penetration.

9. A(n) _Tubular/Slip Joint_ is a pipe size that is smaller in diameter and has a thinner wall thickness than pipe used for drainage piping.

10. A seal that ensures water does not leak and sewer gases do not escape when connecting the toilet to the drainage system is a(n) _Wax Seal_.

Name: _____   Date: _____

Job Sheet 1: Common Tools and Items Required to Install a Toilet

- Upon completion of this job sheet, you should be able to identify the tools needed to install a toilet.
- This job sheet is a checklist of the basic tools used to install a toilet.

Qty.	Tool or Item	Present	Condition
1	Tubing cutter based on type of tank supply used (copper or flexible tubing)	☐	_____
2	Adjustable wrenches or compatible-sized open-end wrenches	☐	_____
1	Adjustable pliers (channel lock or water pump type)	☐	_____
1	Flat-head type screwdriver	☐	_____
1	Miniature hacksaw to trim bolts (standard hacksaw blade only will suffice)	☐	_____
1	Tube of caulking that matches the toilet color (if required)	☐	_____
1	Rag to clean work area	☐	_____

INSTRUCTOR'S RESPONSE

CHAPTER 15   Fixture and Equipment Installation   145

Name: _____   Date: _____

Job Sheet 2: Common Tools and Items Required to Install a Cultured Marble Lavatory

- Upon completion of this job sheet, you should be able to identify the tools needed to install a cultured marble lavatory.
- This job sheet is a checklist of the basic tools used to install a cultured marble lavatory.

Qty.	Tool or Item	Present	Condition
1	Tubing cutter	☐	_____
2	Adjustable wrenches or compatible-sized open-end wrenches	☐	_____
1	Adjustable pliers	☐	_____
1	PVC saw to cut the rough-in pipe	☐	_____
1	Can of purple primer and glue to connect trap adapter to rough-in pipe	☐	_____
1	Copper tubing cutter to cut 1-¼" threaded tailpiece	☐	_____
1	Basin wrench to install faucet and lavatory supply tubing	☐	_____
1	Can of pipe dope or roll of Teflon tape for threads on pop-up assembly items	☐	_____
1	Rag to clean work area	☐	_____

INSTRUCTOR'S RESPONSE

Name: _____   Date: _____

Job Sheet 3: Common Tools and Items Required to Install a Compression Stop

- Upon completion of this job sheet, you should be able to identify the tools needed to install a compression stop.
- This job sheet is a checklist of the basic tools used to install a compression stop.

Qty.	Tool or Item	Present	Condition
1	Tubing cutter based on type of stub-out pipe (copper or flexible tubing)	☐	_____
2	Adjustable wrenches or compatible-sized open-end wrenches	☐	_____
1	Bucket or container to capture water exiting cut pipe	☐	_____
1	Rag to clean spilled water or debris	☐	_____

INSTRUCTOR'S RESPONSE

Plumbing Repairs and Troubleshooting

CHAPTER 16

OBJECTIVES

Upon completion of this chapter, you will be able to:

Knowledge-Based
- Know basic and safe troubleshooting approaches.
- Understand that a troubleshooting approach is based on a specific product or system.
- Recognize that a manufacturer warranty applies to the replacement of a defective product.
- Know that a manufacturer of a specific product provides repair information based on that product.
- Understand basic water heater, well pump, and toilet repair approaches.

KEYWORDS

circuit
circuit breaker
foot valve
heating element
high limit
meter
ohm (Ω)
submersible pump
volt (V)
watts

INTRODUCTION

Once a plumbing system has been installed, it must be maintained. Although many of the large plumbing contractors have plumbing crews that do installation and other crews that do repair and maintenance, it is still a good idea to be fluent in both the installation and the repair of plumbing systems and equipment.

SAFETY

- Because many plumbing repairs involve working with water, electricity, and gas, a plumber must be qualified, and possibly licensed, to repair certain aspects of a plumbing system.
- Extreme caution should be taken when repairing systems that require a plumber to work with electricity, natural gas, and propane.

ELECTRIC WATER HEATERS

- Most residential electric water heaters are 240-volt, non-simultaneous with two heating elements.
- A non-simultaneous heating cycle operates only one heating element at a time.

- The identical elements are positioned in such a way that it is referred to as the upper and lower element.
- The temperature-regulating components are the thermostat and a high-limit device.

High-Limit Devices

- A wire providing electricity to a water heater is known as a leg.
- The two different wires that connect to the high-limit device are identified as line voltage one and two.
- The screws that secure the wire connection to the electrical devices are known as terminals or posts.
- Each wire provides 120 volts of electricity to the heating element, 120 volts per side.

Upper Thermostat

- Most safety standards do not allow a plumber to set the temperature above 120°F.
- The upper thermostat of a seven-pole design has only three posts.
- This three-post design combined with the four posts of the high-limit switch is how the seven-pole design gets its recognition.
- The upper thermostat has a temperature setting feature that is typically either identified alphabetically from A through D or identified as warm, hot, and very hot.

Lower Thermostat

- The lower thermostat has two posts (numbered 1 and 2).
- It also has a temperature setting feature that is similar in design to the upper thermostat.
- The thermostats are secured in place and held against the surface of the tank with a retainer clip.

Heating Elements

- Residential electric water heating elements are typically a screw-in type, but bolt-in types are used for certain water heater designs.
- The electrical and wattage rating of the element is indicated by a manufacturer on the element so its identification is available from the exterior.

Chapter Review Questions and Exercises

COMPLETION

1. A(n) _____ is a unit of measurement describing the resistance of the flow of electrical current.

2. An electrical wiring system that runs from a power source to an equipment or a device is a(n) _____.

3. A(n) _____ is a pump motor and impeller assembly that is submersed in a well below ground to provide water to a piping system.

4. A check valve device installed at the bottom of a vertical drop pipe in a well that serves an aboveground pump to ensure water remains in the pipe after a pumping cycle is a(n) _____.

5. A(n) _____ is an immersed heating device that heats water in a storage tank when energized with electricity.

6. An electrical testing tool used to test voltage, amperage, and continuity of an electrical system is called a(n) _____.

7. A(n) _____ is a safety device in a water heater that disconnects an energy source when the water temperature approaches unsafe levels.

8. _____ is the measure of electromotive force (EMF) and often described as electrical pressure, but a plumber relates it to the amount measured on a voltage meter, such as 120 volts or 240 volts.

9. A device that disconnects (isolates) an electrical circuit is a(n) _____.

10. _____ is a measure of electrical power or the true power in a circuit, but a plumber relates it to a heating element of an electric water heater element, such as 4500 watts.

Name: _____   Date: _____

Job Sheet 1: Water Heater Identification

- Upon completion of this job sheet, you should be able to identify the major components of a water heater.

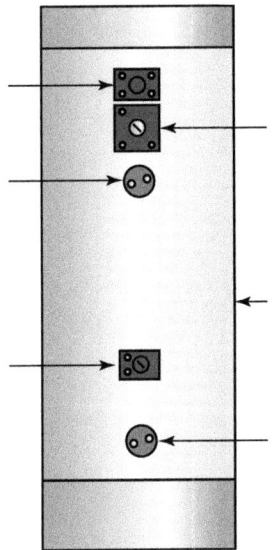

INSTRUCTOR'S RESPONSE

Name: _____   Date: _____

Job Sheet 2: Purging a System

- Upon completion of this job sheet, you should be able to explain the local plumbing code requirements on purging.

After an installation or often a repair, air must be purged from a gas piping system to allow for ignition of the equipment, such as a water heater. Purging requires a plumber to use piping or hoses temporarily connected to the piping system and routing the hose or temporary pipe to a safe exterior area.

Research the local plumbing codes in your area to determine the requirements for purging the air from a system.

INSTRUCTOR'S RESPONSE

Name: _____   Date: _____

Job Sheet 3: Electrical Wiring Requirements for a Plumber

- Upon completion of this job sheet, you should be able to explain the local code requirements for a plumber to perform electrical work.

Your local code may not allow a plumber to perform electrical work without proper certification.

Research the local codes in your area to determine the requirements for plumbers to perform electrical work without proper certification.

INSTRUCTOR'S RESPONSE

Name: _____ Date: _____

Job Sheet 4: Disassembling a Gas Regulator

- Upon completion of this job sheet, you should be able to explain the local code requirements for a plumber to dissemble a gas regulator.

Your local code may not allow a plumber to disassemble a gas regulator; most have security screws that require a special tool to remove the cover.

Research the local codes in your area to determine the requirements for plumbers to disassemble a gas regulator.

INSTRUCTOR'S RESPONSE

CHAPTER 17

Hydronic Heat

OBJECTIVES
Upon completion of this chapter, you will be able to:

Knowledge-Based
- Explain the concept of hydronic heating.
- List the three most commonly used heat sources in boilers.
- Describe basic boiler construction.
- Identify component parts of a boiler.
- Explain the operation of a boiler.
- Describe various components that maintain the desired water temperature in a boiler.
- Explain the difference between a one-pipe and a two-pipe hot water system.
- Discuss the difference between direct-return systems and reverse return systems.
- Describe the operation and function of centrifugal pumps.
- Explain the function of boiler controls and safety devices.
- Explain the function of an expansion tank.
- Explain the point of no pressure change.
- Explain primary-secondary pumping.
- Explain the concept of "zoning."
- Explain how a radiant heating system creates comfort.
- Explain how a radiant heating system operates.

Skill-Based
- Check the pressure in an expansion tank.
- Install a boiler.
- Service boilers.

KEYWORDS

air cushion
air separator
air vent
aquastat
automatic air vent
balancing valve
baseboard sections
boiler
boiler/feed water valve

centrifugal pump
circulator
closed loop
compression tank
diaphragm-type expansion tank
direct return
diverter tee
expansion tank

feet of head
flow check valve
flow-control valve
hydronics
indirect water storage tank
manifold station
manual air vent
monoflo tee
one-pipe hot water

outdoor reset
PEX tubing
point of no pressure change
polyethylene tubing
pressure-reducing valve
pressure relief valve
radiant heat system

radiator
relief valve
reverse return system
standard expansion tank
steam boiler
tankless hot water heating coil
terminal unit

two-pipe system
volume factor
volute
zone valve
zoning

INTRODUCTION

Hydronic heating is a cross-discipline field in which both plumbers and HVAC technicians currently design, install, and maintain. Although the system is used primarily for providing heating to a structure, it does incorporate water, piping, and various other devices and fittings currently used by the plumbing trade. Therefore, a good understanding of this field is beneficial to a plumbing technician.

THEORY OF HYDRONIC HEATING SYSTEMS

- Hydronic systems rely on circulating water or steam to deliver heat to the remote locations where heating is desired.
- Theory of hydronic heating systems.

THE HEAT SOURCE

- Heat energy is transferred from the heat source to the water in the boiler.
- Cast iron boilers are the most commonly found boilers in residential applications.
  – Residential cast iron boilers typically hold between 10 and 15 gallons of water.

AQUASTAT

- The aquastat is a temperature-sensing switch that is responsible for cycling the boiler on and off to keep the water in the boiler close to the desired temperature.

RESET

- The reset thermostat senses the outdoor temperature and adjusts the water temperature in the boiler.
  – As the outdoor temperature increases, the water temperature in the boiler will be reduced.

LOW WATER CUTOFF

- The low water cutoff is responsible for de-energizing the boiler in the event the water level in the system falls below the desired level.

CHAPTER 17 Hydronic Heat

EXPANSION TANKS
- There are two types of expansion tanks:
  - Standard expansion tank: a large tank located above the boiler.
  - Diaphragm-type expansion tank: divided into two sections separated by a rubber, semi-permeable membrane.
- One side of the tank contains air and the other side is open to the water circuit.

CENTRIFUGAL PUMPS
- The pumps used to move water through hydronic systems are often called circulators.
- The centrifugal pump is made up of a motor, a linkage, and an impeller.

AIR VENT AND AIR SEPARATORS
- One of the biggest enemies of a hot water hydronic heating system is air.
- The air separator is designed to remove smaller air bubbles from the system.

CHAPTER REVIEW QUESTIONS AND EXERCISES

COMPLETION

1. _____ is a spring-loaded valve that opens when the pressure in a hydronic system exceeds the rating of the valve.

2. _____ is the air above the semi-permeable membrane in an expansion tank.

3. _____ is a fitting that automatically removes air from a hydronic heating system.

4. _____ is a valve that reduces the pressure entering the structure to the pressure required by the hydronic system.

5. _____ is a pump that moves water through a piping circuit by means of centrifugal force.

6. _____ is a device that is used to remove air from a hydronic system. Air vents can be manual or automatic devices.

7. _____ is a device that removes air from the system.

8. _____ is a fitting used to remove air, either manually or automatically, from a hydronic system.

9. _____ is a valve that prevents backward and gravity circulation through loops.

10. _____ is a manually controlled valve used to increase the resistance and reduce water flow through a given branch circuit.

11. _____ is a system-piping component that provides additional space for expanding water to occupy.

12. _____ is an electrical component that opens and closes its contacts to energize and de-energize electric circuits in response to the temperature sensed by the device.

13. _____ is a term used to describe a system that is closed or isolated from the atmosphere.

14. _____ is a hydronic piping configuration that utilizes a main hot water loop and diverter tees to connect the terminal units to the system.

15. _____ is the term used for heat emitters or terminal units that transfer the majority of their heat to the occupied space by radiation.

16. _____ is a heating system that attempts to regulate the heat loss of the individual as opposed to the rate of heat loss of the structure.

17. _____ is the term used to describe a radiator or section of baseboard.

18. _____ is the process of dividing the structure into separate areas, each of which has its own means to regulate the temperature in the space.

19. _____ is a hydronic piping configuration that utilizes one pipe as the supply and one pipe as the return. It can be configured as a direct return or a reverse return.

20. _____ is a thermostatically controlled valve that opens and closes to regulate the flow of hot water to the terminal units in the occupied space.

Name: _____   Date: _____

Job Sheet 1: Installing a Hydronic System

- Upon completion of this job sheet, you should be able to install a hydronic system.

Installing and putting a hot water hydronic system into operation involves a number of steps.

Step	Complete
Setting and installing the boiler	☐
Installing the piping	☐
Wiring the unit	☐
Filling the system	☐
Starting up the system	☐

INSTRUCTOR'S RESPONSE

Name: _____ Date: _____

Job Sheet 2: Installing the Boiler

- Upon completion of this job sheet, you should be able to install a boiler in a hydronic system.

When setting and installing the boiler, keep the following steps in mind.

Step	Complete
Make certain the boiler is level.	☐
Locate boiler as close as possible to the chimney.	☐
Make certain the flue piping is installed according to the manufacturer's literature.	☐
Make certain the amount of air introduced to the area around the boiler is sufficient for combustion and dilution.	☐
Make certain all packing material has been removed from the boiler.	☐
Install the pressure relief in the tap specified by the manufacturer.	☐
Provide enough clearance around the boiler for future service.	☐
Make certain that a disconnect switch is located on or very close (within 2 feet) to the boiler.	☐
Check the combustion of the boiler on initial start-up.	☐
Properly set the pressure-reducing valve prior to filling.	☐
Properly check and adjust the air pressure in the expansion tank prior to filling the system.	☐

INSTRUCTOR'S RESPONSE

Name: _____   Date: _____

Job Sheet 3: Installing the Piping

- Upon completion of this job sheet, you should be able to install the piping of a hydronic system.

When installing the piping circuit, keep the following steps in mind.

Step	Complete
Keep vertical piping vertical and keep horizontal piping horizontal.	☐
In addition to proper alignment of fittings and components, customers will appreciate a neat looking job, and so will you and your boss.	☐
Always try to keep the number of fittings to a minimum.	☐
Excess fittings increase the cost of the job and the chance of water leakage.	☐
For the sake of neatness and ease in wiring, group similar components, such as circulators or zone valves, together.	☐
Also keep them at the same height and have them all pointing in the same direction. Neatness counts!	☐
Use lots of valves.	☐
Although this increases the cost of the materials for the job, time and money will be saved when it comes time to performing service in the future.	☐
Valves placed before and after a circulator, for example, will save the technician the trouble of having to drain the entire system to change a defective impeller.	☐
Properly support the piping and components to avoid sagging.	☐
Test all tubing for leaks (radiant heat systems) prior to pouring concrete!	☐

INSTRUCTOR'S RESPONSE

Name: _____ Date: _____

Job Sheet 4: Filling the System

- Upon completion of this job sheet, you should be able to fill a hydronic system.

Step	Complete
Determine how much water pressure is needed to lift water to the highest point in the system. One pound-force per square inch gauge of water pressure will lift water 2.3 feet.	☐
Once the water pressure has been determined, add 3 psi pressure.	☐
Check the air pressure inside the compression tank.	☐
The air pressure should equal the pressure you plan to use on the waterside of the diaphragm.	☐
When starting up the system, keep the following in mind:	
Check the operation of all safety components and accessories on initial start-up.	☐
Test the operation of all circulators, zone valves, thermostats, etc.	☐
Make certain the thermostats control the operation of the correct zone valves and circulators.	☐
Make certain that the boiler cycles on and off on the aquastat.	☐
Test the operation of all safety controls.	☐
Test the combustion efficiency of the boiler.	☐
Determine carbon dioxide and carbon monoxide levels and compare them to acceptable levels.	☐

INSTRUCTOR'S RESPONSE